MOLECULAR BIOLOGY INTELLIGENCE UNIT

RNA Binding Proteins

Zdravko J. Lorković, BSc, PhD
Gregor Mendel Institute for Molecular Plant Research
Vienna, Austria
and
Department of Molecular Biology
University of Zagreb
Zagreb, Croatia

CRC Press
Taylor & Francis Group
Boca Raton London New York

CRC Press is an imprint of the
Taylor & Francis Group, an **informa** business

RNA Binding Proteins

Molecular Biology Intelligence Unit

First published 2012 by Landes Bioscience

Published 2018 by CRC Press
Taylor & Francis Group
6000 Broken Sound Parkway NW, Suite 300
Boca Raton, FL 33487-2742

© 2012 by Taylor & Francis Group, LLC
CRC Press is an imprint of Taylor & Francis Group, an Informa business

First issued in paperback 2019

No claim to original U.S. Government works

ISBN 13: 978-0-367-44591-1 (pbk)
ISBN 13: 978-1-58706-656-6 (hbk)

Visit the Taylor & Francis Web site at
http://www.taylorandfrancis.com

and the CRC Press Web site at
http://www.crcpress.com

Library of Congress Cataloging-in-Publication Data

Lorkovic, Zdravko J., 1967-
RNA binding proteins / Zdravko J. Lorkovic.
   p. ; cm. -- (Molecular biology intelligence unit)
Includes bibliographical references and index.
ISBN 978-1-58706-656-6 (alk. paper)
I. Title. II. Series: Molecular biology intelligence unit (Unnumbered : 2003)
[DNLM: 1. RNA-Binding Proteins. QU 55.2]

579.2'5--dc23
          2012021396

About the Editor...

ZDRAVKO J. LORKOVIĆ is a senior scientist in the group of Marjori and Antonius Matzke at Gregor Mendel Institute for Molecular Plant Research in Vienna, Austria and is Associate Professor at the Department of Molecular Biology, Faculty of Science at the University of Zagreb, Croatia. Zdravko received a BSc degree from the University of Zagreb (Croatia) and a PhD in Plant Biology from the Ludwig-Maximilans-University in Munich (Germany, 1997). He worked on plant RNA binding proteins involved in intron recognition and splicing and on fission yeast cyclophiln regulating RNA polymerase II transcription. Currently, his major topic is RNA dependent DNA methylation in Arabidopsis. Research interests include RNA polymerase II transcription and its coupling with pre-mRNA processing, plant RNA binding proteins and regulation and function of plant RNA polymerases IV and V.

CONTENTS

EDITOR

Zdravko J. Lorković
Gregor Mendel Institute for Molecular Plant Research
Vienna, Austria
and
Department of Molecular Biology
University of Zagreb
Zagreb, Croatia

CONTRIBUTORS

Frédéric H.-T. Allain
Institute for Molecular Biology
  and Biophysics
ETH Zürich
Zürich, Switzerland
Chapter 9

Jérôme Barbier
Département de Microbiologie
  et d'Infectiologie
Faculté de Médecine et des Sciences
  de la Santé
Université de Sherbrooke
Sherbrooke, Québec, Canada
Chapter 1

Benoit Chabot
Département de Microbiologie
  et d'Infectiologie
Faculté de Médecine et des Sciences
  de la Santé
Université de Sherbrooke
Sherbrooke, Québec, Canada
Chapter 1

Antoine Cléry
Institute for Molecular Biology
  and Biophysics
ETH Zürich
Zürich, Switzerland
Chapter 9

Andrew J. Crofts
International Liberal Arts Program
Akita International University
Akita, Japan
Chapter 6

Naoko Crofts
International Liberal Arts Program
Akita International University
Akita, Japan
Chapter 6

Ralf Dahm
Department of Biology
University of Padova
Padova, Italy
Chapter 4

Kelly A. Doroshenk
Institute of Biological Chemistry
Washington State University
Pullman, Washington, USA
Chapter 6

Masako Fukuda
Faculty of Agriculture
Kyushu University
Fukuoka, Japan
Chapter 6

Katherine S. Godin
MRC Laboratory of Molecular Biology
Cambridge, UK
Chapter 3

Klemens J. Hertel
Department of Microbiology
and Molecular Genetics
University of California
Irvine, California, USA
Chapter 2

Michael F. Jantsch
Department of Chromosome Biology
Max F. Perutz Laboratories
University of Vienna
Vienna, Austria
Chapter 7

Hunseung Kang
Department of Plant Biotechnology
College of Agriculture and Life Sciences
Chonnam National University
Gwangju, Korea
Chapter 8

Toshihiro Kumamaru
Faculty of Agriculture
Kyushu University
Fukuoka, Japan
Chapter 6

Kyung Jin Kwak
Department of Plant Biotechnology
College of Agriculture and Life Sciences
Chonnam National University
Gwangju, Korea
Chapter 8

Paolo Macchi
Centre for Integrative Biology
University of Trento
Trento, Italy
Chapter 4

Laetitia Michelle
Département de Microbiologie
et d'Infectiologie
Faculté de Médecine et des Sciences
de la Santé
Université de Sherbrooke
Sherbrooke, Québec, Canada
Chapter 1

Robert T. Morris
School of Molecular Biosciences
Washington State University
Pullman, Washington, USA
Chapter 6

William F. Mueller
Department of Microbiology
and Molecular Genetics
University of California
Irvine, California, USA
Chapter 2

Dierk Niessing
Helmholtz Zentrum München
Institute for Structural Biology
Neuherberg, Germany
and
Gene Center Munich
Munich University
Munich, Germany
Chapter 5

Thomas W. Okita
Institute of Biological Chemistry
Washington State University
Pullman, Washington, USA
Chapter 6

Alessandro Quattrone
Centre for Integrative Biology
University of Trento
Trento, Italy
Chapter 4

Boris Rogelj
MRC Centre for Neurodegeneration
Research
Institute of Psychiatry
King's College London
London, UK
and
Department of Biotechnology
Jozef Stefan Institute
Ljubljana, Slovenia
Chapter 3

Hikaru Satoh
Faculty of Agriculture
Kyushu University
Fukuoka, Japan
Chapter 6

Mio Satoh-Cruz
Institute of Biological Chemistry
Washington State University
Pullman, Washington, USA
Chapter 6

Christopher E. Shaw
MRC Centre for Neurodegeneration
  Research
Institute of Psychiatry
King's College London
London, UK
Chapter 3

Jernej Ule
MRC Laboratory of Molecular Biology
Cambridge, UK
Chapter 3

Cornelia Vesely
Department of Chromosome Biology
Max F. Perutz Laboratories
University of Vienna
Vienna, Austria
Chapter 7

Haruhiko Washida
Laboratory of Plant Molecular Genetics
Nara Institute of Science and Technology
Ikoma, Japan
Chapter 6

Yongil Yang
Institute of Biological Chemistry
Washington State University
Pullman, Washington, USA
Chapter 6

PREFACE

Now it is clear that post-transcriptional control of gene expression in eukaryotes is as important as the control of transcription. In all aspects of post-transcriptional gene regulation a crucial role for RNA binding proteins has been documented. In fact, RNAP II transcripts are accompanied by the RBPs from the start of the transcription until they are degraded in the cytoplasm.

The most abundant nuclear RNA-binding proteins in human cells are collectively termed heterogeneous nuclear ribonucleoproteins (hnRNPs), according to their association with nascent RNA polymerase II transcripts. Although they were originally thought to be packaging molecules protecting RNA from degradation it is clear now that they play active roles in virtually all aspects of gene regulation at the post-transcriptional level. Chapter 1 by Benoit Chabot and colleagues provides an introduction to extended family of hnRNPs, some details of how they contribute to generic and alternative splicing and how misregulation of their functions can lead to many pathological states in humans.

Chapter 2 by William Mueller and Klemens Hertel address SR protein family and diverse mechanisms of how these proteins regulate alternative splicing.

In addition to well-defined RNA binding domains most RNA binding proteins contain auxiliary domains of unusual amino acids composition. The function-structure relationship of glycine-rich domains in several RNA binding proteins and their importance in human diseases are discussed in Chapter 3 by Jernej Ule and colleagues.

The following three chapters address the localization of mRNA, an emerging aspect of gene expression control. In Chapter 4 Paolo Macchi and colleagues discuss mechanisms of RNA localization in vertebrate nervous system. This is followed by Chapter 5 where Dierk Niessing summarizes new insights of how RNA binding proteins contribute to RNA localization in a single-celled eukaryote *Saccharomyces cerevisae* as well as in filamentous fungi *Candida albicans* and *Ustilago maydis*. Then in Chapter 6, Thomas Okita and colleagues describe relatively new phenomenon of RNA localization in plant cells and its impact on storage protein synthesis.

In Chapter 7, Michael Jantsch and Cornelia Vesely describe the mechanism of RNA editing and involvement of diverse RNA binding proteins.

Function of RNA binding proteins in plants is still largely unknown. Hunseung Kang and Kyung Jin Kwak discuss recent data on impacts of RNA binding proteins on plant response to abiotic stresses in Chapter 8.

Finally, Chapter 9 by Antoine Cléry and Frédéric Allain addresses the structural analysis of RNA binding proteins which revealed important insights into how different RNA binding domains contribute to recognition of diverse and specific RNA sequences.

I wish to thank to all the contributors for providing clear, informative and well-illustrated chapters, which although covering only a part of post-transcriptional regulation of gene expression, provide a stimulating overview of RNA binding proteins and their highly diverse and versatile function.

Zdravko J. Lorković, BSc, PhD
Gregor Mendel Institute for Molecular Plant Research
Vienna, Austria
and
Department of Molecular Biology
University of Zagreb
Zagreb, Croatia

CHAPTER 1

hnRNP and hnRNP-Like Proteins in Splicing Control:
Molecular Mechanisms and Implication in Human Pathologies

Laetitia Michelle, Jérôme Barbier and Benoit Chabot*

Abstract

The heterogeneous nuclear ribonucleoprotein (hnRNP) family includes a diverse group of RNA binding proteins to which we are affiliating here other structurally and functionally related proteins (e.g., Nova, Sam68, ESRP, Fox, TDP-43, Hu, CUG-BP, MBNL and TIA proteins). These hnRNP and hnRNP-like proteins make important contributions to protein diversity and activity by modulating the alternative splicing of a large repertoire of pre-mRNAs. They achieve this function through a variety of molecular strategies ranging from directly preventing the recognition of splice sites, antagonizing or helping the assembly of positive regulatory complexes, interfering with spliceosome assembly and changing the conformation of pre-mRNAs. In addition to regulating key splicing events, defects in the expression of these proteins are now being documented for a growing number of human diseases including cancer. hnRNP and hnRNP-like proteins therefore represent a group of proteins whose roles in the control of splicing pervade all areas of biology and human health.

Introduction

Heterogeneous nuclear ribonucleoproteins (hnRNPs) form a family of abundant RNA binding proteins associating with pre-messenger RNAs.[1-3] Membership to the family was based initially on ultraviolet crosslinking and immunoprecipitation assays, and included a defined set of approximately 25 polypeptides.[3] In addition to their role in pre-messenger RNA (pre-mRNA) splicing which will be covered below, hnRNP proteins participate in a wide variety of processes including transcription regulation,[4-7] maintenance of telomere length and telomerase activity,[8,9] nuclear retention,[10] mRNA stability[11] and 3'-end processing.[12] Most hnRNPs shuttle between the nucleus and the cytoplasm,[13] providing an opportunity for these proteins to transport mRNAs, as well as control translation.[14,15] The number of hnRNP proteins is now increasing because splice variants and paralogues of the original members are being identified. Moreover, while newly described proteins are regularly associated with the hnRNP family based on structural features, others remain excluded. It would therefore be justified to revisit the rules that determine membership to this group of RNA binding proteins. Here, based on structural and functional considerations we have extended the hnRNP family by including Nova, Sam68 and QKI, as well as MBNL and CUG-BP family members, TDP-43, TIA1, TIAR, HuD, HuR, Fox2 and ESRP proteins.

*Département de Microbiologie et d'Infectiologie, Faculté de Médecine et des Sciences de la Santé, Université de Sherbrooke, Sherbrooke, Québec, Canada.
Corresponding Author: Benoit Chabot—benoit.chabot@usherbrooke.ca

RNA Binding Proteins, edited by Zdravko J. Lorković.
©2012 Landes Bioscience.

We begin by reviewing some of the structural features of hnRNP and hnRNP-like proteins and discussing the molecular mechanisms used by these proteins to control splicing decisions. We will conclude by presenting the clearest examples supporting a role for this extended family of hnRNP proteins in human diseases.

Structural Diversity of hnRNP and hnRNP-Like Proteins

The classical hnRNP proteins are designated hnRNP A to hnRNP U. Many splice variants and newly identified proteins have been added to the initial list based on structural similarities. We have excluded from the hnRNP-like group proteins any protein that contains an RS domain reminiscent of SR proteins (see Chapter 2 by Mueller and Hertel). The overall structure of hnRNP and hnRNP-like proteins is modular and each member contains one or more domains that mediate RNA binding, protein-protein interaction or cellular localization.

The most common RNA binding domain is the RNA recognition motif (RRM) found in hnRNP proteins A1, D, L, Q, RBMY, PTB, and also in the following hnRNP-like splicing regulators: Raver1, TDP-43, TIA1, TIAR1, HuR, HuD, Fox2, CUG-BP and ESRP proteins (Fig. 1). RRM domains are also found in other classes of RNA binding proteins (e.g., SR, SR-related and several snRNP-associated proteins). The RRM consists of approximately 90 amino acids (aa) forming four strands and two helices arranged in an α/β sandwich (see Chapter 9 by Cléry and Allain for more details), with a third helix contributing to RNA binding in the case of the quasi-RRM (qRRM) of hnRNP F and H.[16] One RRM in PTB can also engage in protein/protein interaction.[17] In hnRNP C, the RRM plays a minimal role in RNA binding, RNA affinity being mostly determined by a highly basic domain that precedes the leucine zipper motif.[18-21] hnRNP U binds RNA through its glycine-rich domain.[22] hnRNP P2 (also known as TLS/FUS) possesses one RRM and several RGG motifs, but RNA binding appears to be mediated by a zinc finger domain.[23] Similarly, MBNL proteins possess four zinc finger domains organized in tandem pairs and responsible for RNA recognition.[24] The KH domain is an RNA binding domain that was first described in hnRNP K (Fig. 2). It is approximately 70 aa long and contains conserved octapeptide repeats of Ile-Gly-X-X-Gly-X-X-Ile (X being any aa). KH domains are also found in hnRNP E and the hnRNP-like Nova proteins, Sam68 and QKI. Through these various RNA binding motifs, hnRNP and hnRNP-like proteins will recognize and bind sequences with a range of specificities (Figs. 1 and 2).

Glycine-rich domain (GRD) and the RGG domains are often found in hnRNP and hnRNP-like proteins (Figs. 1 and 2). They have been associated with RNA binding but are mostly known for their role in protein-protein interactions. hnRNP C contains a nuclear retention signal (NRS) that prevents it from shuttling to the cytoplasm like the other hnRNP proteins. hnRNP H proteins

Figure 1, viewed on following page. Schematic representation of the RRM domain-containing hnRNP and hnRNP-like proteins. Different isoforms produced by alternative splicing are also indicated. A1B is generated when alternative exon 7B is included, adding 52 aa in the human protein at the C-terminal glycine-rich region, while B1 contains an additional 12 aa stretch near the amino terminus. hnRNP C1 and C2 are also generated by alternative splicing. hnRNP D proteins include four different splice variants: p37 (also known as AUF1 or D01), p40 (D02), p42 (D1) and p45 (D2). D02 and D1 result from the inclusion of alternative exons 2 and 7, respectively. D2 is generated when both exons 2 and 7 are included. Six isoforms of hnRNP 2H9 are produced by alternative splicing and 4 are represented here. PTB is alternatively spliced to produce a variety of isoforms that differ in the length of the interdomain linker. hnRNP Q comes in three versions: Q1, Q2 and Q3. hnRNP R is 83% identical to Q3 and has 2 different splicing isoforms. bZLM: basic leucine zipper-like domain; G-rich: glycine-rich domain; GY-rich: glycine/tyrosine-rich domain; GYR-rich: glycine/tyrosine/arginine-rich domain; NRS: nuclear retention signal; M9: nucleocytoplasmic export signal; Q-rich: glutamine-rich domain; QN-rich: glutamine/asparagine-rich domain; RRM: RNA recognition motif; qRRM: quasi-RNA recognition motif; RGG: arginine-glycine-glycine motif-containing domain; SAF: scaffold-associated region-specific bipartite DNA binding domain; SPRY: SP1a and ryanodine receptor homology domain; SRGY: serine/arginine/glycine/tyrosine-rich domain; Zn: zinc finger domain; Y = C or U (pyrimidines); R = A or G (purines).

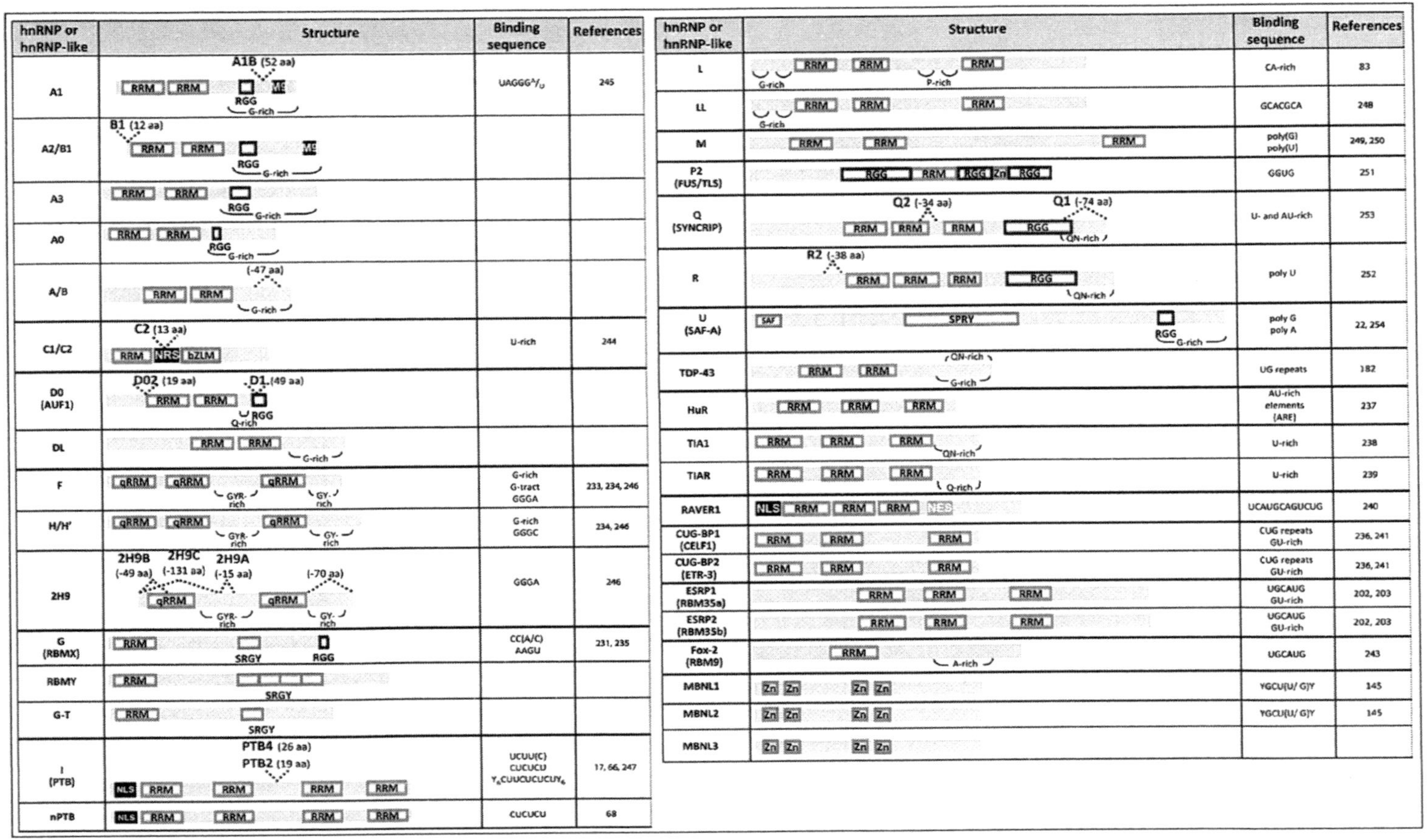

hnRNP or hnRNP-like	Structure	Binding sequence	References
A1	RRM RRM (A1B 52 aa) G-rich, RGG	UAGGG/$_U$	245
A2/B1	RRM RRM (B1 12 aa) RGG G-rich		
A3	RRM RRM RGG G-rich		
A0	RRM RRM RGG G-rich		
A/B	RRM RRM (-47 aa) G-rich	U-rich	244
C1/C2	RRM NLS bZLM (C2 13 aa)		
D0 (AUF1)	RRM RRM (D02 19 aa)(D1 49 aa) G-rich RGG		
DL	RRM RRM G-rich		
F	qRRM qRRM GY-rich qRRM GY-rich	G-rich, G-tract, GGGA	233, 234, 246
H/H'	qRRM qRRM GY-rich qRRM GY-rich	G-rich, GGGC	234, 246
2H9	qRRM (2H9B -49 aa)(2H9C -131 aa)(2H9A -15 aa) qRRM GY-rich (-70 aa) qRRM	GGGA	246
G (RBMX)	RRM SRGY RGG	CC(A/C) AAGU	231, 235
RBMY	RRM SRGY		
G-T	RRM SRGY		
I (PTB)	RRM RRM (PTB4 26 aa)(PTB2 19 aa) NLS RRM RRM	UCUU(C) CUCUCU Y$_n$CU$_s$UGCUCUCUU$_n$	17, 66, 247
nPTB	NLS RRM RRM RRM RRM	CUCUCU	68

hnRNP or hnRNP-like	Structure	Binding sequence	References
L	RRM RRM P-rich RRM	CA-rich	83
LL	RRM RRM RRM	GCACCCA	248
M	RRM RRM RRM	poly(G) poly(U)	249, 250
P2 (FUS/TLS)	RGG RRM RGG Zn RGG	GGUG	251
Q (SYNCRIP)	RRM RRM RRM (Q2 -34 aa)(Q1 -74 aa) RGG QN-rich	U- and AU-rich	253
R	RRM RRM RRM (R2 -38 aa) RGG QN-rich	poly U	252
U (SAF-A)	SAP SPRY QN-rich	poly G poly A	22, 254
TDP-43	RRM RRM G-rich	UG repeats	182
HuR	RRM RRM RRM	AU-rich elements (ARE)	237
TIA1	RRM RRM RRM QN-rich	U-rich	238
TIAR	RRM RRM RRM Q-rich	U-rich	239
RAVER1	NLS RRM RRM RRM NES	UCAUGCAGUCUG	240
CUG-BP1 (CELF1)	RRM RRM RRM	CUG repeats GU-rich	236, 241
CUG-BP2 (ETR-3)	RRM RRM RRM	CUG repeats GU-rich	236, 241
ESRP1 (RBM35a)	RRM RRM RRM	UGCAUG GU-rich	202, 203
ESRP2 (RBM35b)	RRM RRM RRM	UGCAUG GU-rich	202, 203
Fox-2 (RBM9)	RRM A-rich	UGCAUG	243
MBNL1	Zn Zn Zn Zn	YGCU(U/ G)Y	145
MBNL2	Zn Zn Zn Zn	YGCU(U/ G)Y	145
MBNL3	Zn Zn Zn Zn		

Figure 1. Please see the figure legend on previous page.

hnRNP or hnRNP-like	Structure	RNA binding site	References
E1/E2	KH KH KH	poly C	255
K	KH...KH KH KH KI KNS KH J (24 aa) RGG	poly C $AUC_{3/4}(^U/_A)(^A/_U)$	256
Nova-1	KH KH KH	Clusters of YCAY	257
Nova-2	KH KH KH	GAGUCAU	258
Sam68	P-rich P-rich KH P-rich P-rich P-rich NLS RGG Y-rich	UAAA, UUUUUU	259, 260
QKI	KH RGG	ACUAAY[...]UAAY	261
FMR1	KH KH RGG	Poly G	262

Figure 2. Schematic representation of the KH domain-containing hnRNP and hnRNP-like proteins. KH: hnRNP K homology domain; KI: K interactive domain; KNS: K nuclear shuttle domain; P-rich: proline-rich domain; Y-rich: tyrosine-rich domain; NLS: nuclear localization signal; RGG: arginine-glycine-glycine motif-containing domain; Y = C or U (pyrimidines).

contain a C-terminal glycine- and tyrosine-rich (GY) domain and a region rich in glycine, tyrosine and arginine (GYR) located between qRRM2 and qRRM3.[10,16] hnRNP K possesses a KI domain responsible for interactions with various proteins, a nuclear shuttling domain (KNS) as well as RGG and SH3 binding domains. RBMY has four tandemly repeated motifs (SRGY boxes) also found in SR proteins where they mediate protein-protein interactions. hnRNP A and Q proteins contain a nuclear localization signal (NLS). The larger isoforms of hnRNP Q and R contain a glutamine- and asparagine-rich (QN) C-terminal domain. hnRNP U possesses a SP1a and ryanodine receptor (SPRY) homology domain of unknown function and a scaffold-associated region (SAR)-specific bipartite DNA binding domain (SAF) capable of binding specific DNA sequences.[25] Raver1, like hnRNP A1, possesses a NLS and a nuclear export signal (NES) allowing shuttling between nucleus and cytoplasm.[26] Sam68 contains several proline-rich (P) domains involved in interacting with proteins harboring SH3 and WW domains.

hnRNPs and Generic Splicing

Except for histones and a few other genes, all metazoan genes contain introns that must be removed efficiently and with precision. A typical human gene contains on average 8 exons but multi-exon genes can contain from 2 to 363 exons.[27] While internal exons average 145 nucleotides (nt) in length, introns are usually more than 10 times this size and can be much larger.[28] Constitutive introns separate exons found in the same order in the final mRNA irrespective of tissues, and the act of removing them is called generic splicing. Intron removal and the joining of exons is performed by the spliceosome, a macromolecular ribonucleoprotein complex assembled in a stepwise manner onto the pre-mRNA. Although hnRNP proteins have been identified in various spliceosomal assemblies,[29,30] their roles in these complexes and in generic splicing remain unclear. While a body of work shows that hnRNP proteins can interact with components of the splicing machinery and can even stimulate splicing,[31-34] a supporting role for hnRNP proteins in generic splicing was challenged by the demonstration that the rapid association of hnRNP proteins with model pre-mRNAs forms splicing-incompetent complexes.[35,36] Nevertheless, because the composition of these complexes varies depending on the identity of the pre-mRNA, distinct sets of hnRNP proteins may possibly contribute to the generic splicing of specific introns.

The abundance of putative binding sites for hnRNP A1 and hnRNP H in introns and their greater prevalence in the vicinity of the splice junctions support a role for these proteins in generic splicing.[37,38] Homotypic or heterotypic interactions between hnRNP A1 and H bound near the ends of a large or poorly defined intron may stimulate splicing by helping to bring in closer proximity distant splice sites (Fig. 3A). Consistent with this view, promoting A1 binding in poorly spliced introns stimulated their removal in vitro.[37]

hnRNP and hnRNP-Like Proteins and Alternative Splicing

Transcriptome complexity arises when different combinations of splice sites are used through alternative pre-mRNA splicing to produce distinct mRNAs. Alternative splicing is part of the expression program of more than 90% of all human multi-exon genes[39,40] and is the major source of protein diversity in plants and metazoans. Alternative exons often have suboptimal splice sites, making them prime targets for positive or negative control. Indeed, splice site usage can be improved or repressed by different types of sequence elements, and specific combinations of positive and negative elements can control the selection of splice sites in a wide variety of situations.[41] In general, the activity of enhancer elements is mediated by SR proteins.[42] However, there are now several examples of SR proteins that repress splicing (see Chapter 2 by Mueller and Hertel for more details on SR protein function).[43-45] In contrast, hnRNP proteins have been associated mostly with silencer elements, but here again recent results suggest that hnRNP proteins can also make positive contributions.

Although there are many examples indicating that SR and hnRNP proteins are antagonistic (see below), global studies in *Drosophila* have led to a reconsideration of this principle since only a small fraction of the splicing events controlled by the hnRNP A1 orthologues are also regulated by the SR proteins B52 and dASF/SF2.[46,47] These and other studies have revealed that the depletion

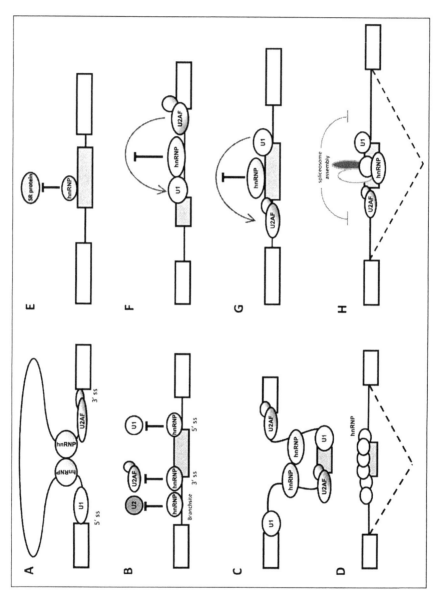

Figure 3. Please see the figure legend on following page.

Figure 3, viewed on previous page. Mechanisms used by hnRNP and hnRNP-like proteins to modulate splicing. A) The generic splicing of a large or poorly defined intron may require interactions between hnRNPs bound to intronic sequence. This process would help intron definition by bringing in closer proximity distant splice sites. B) Direct interference. High affinity binding sites for hnRNP proteins can sterically block the interaction of splicing or regulatory factors with splicing signals or control elements. C) Interactions between RNA-bound hnRNP molecules may loop out portions of a pre-mRNA to promote exon skipping. D) Nucleation of hnRNPs between high affinity sites can mask the splice sites of an exon. E) hnRNP proteins can interfere with exonic enhancers without affecting direct binding of U1 snRNP and U2AF. F) hnRNP proteins can inhibit intron definition when bound to an intron. G) hnRNP proteins can inhibit exon definition when bound to an exon. H) hnRNP proteins within a multiprotein complex can prevent spliceosome assembly. 5'ss: 5' splice site; 3'ss: 3' splice site.

of specific SR, hnRNP and hnRNP-like proteins can promote exon inclusion in some pre-mRNAs but exclusion in others.[48-51] Thus, the positive or negative impact of a regulatory protein on exon inclusion cannot be anticipated by its membership to a family of proteins. As we will see below, the impact of an RNA binding protein on the splicing behavior of a given pre-mRNA depends at least in part on where binding occurs. Moreover, collaboration and/or antagonism between distinct SR, hnRNP and hnRNP-like proteins likely provide the basis for the specificity of alternative splicing.

Molecular Mechanisms of Splicing Control by hnRNP and hnRNP-Like Proteins

Negative Regulation

Direct Interference at Splicing Signals

The splicing impact of a regulatory protein depends on where it binds. In its simplest form, the binding of hnRNP proteins at or near splice sites can interfere with the interaction of components of the splicing machinery, like U2AF, U1 and U2 snRNP, with the pre-mRNA (Fig. 3B and Table 1). The hnRNP-like proteins TDP-43 and Fox-2 also use this strategy to interfere directly with 3'ss/branchsite recognition. Specific cases of steric hindrance are listed in Table 1, but this interference mechanism appears to be used extensively by many hnRNP proteins to control splicing decisions. PTB binding in introns near regulated splice sites promotes the skipping of numerous exons[49,50] (Fig. 4D-E). In contrast, its binding near flanking constitutive splice sites is generally associated with the inclusion of regulated exons, possibly

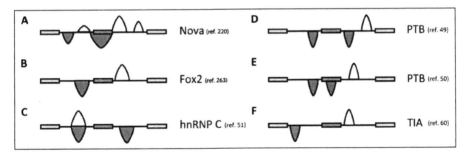

Figure 4. RNA binding maps of hnRNP and hnRNP-like proteins. RNA binding maps relative to a model alternative exon are shown for Nova (panel A), Fox2 (panel B), hnRNP C (panel C), PTB (panels D and E) and TIA1 (panel F). Peaks represent regions of high density binding for the proteins. Peaks that are on top of the pre-mRNA are associated with inclusion of the regulated exon while peaks below the pre-mRNA are associated with skipping of the central exon. For PTB, two maps are shown corresponding to two distincts studies; in panel D, PTB binding close to the 3' and 5' alternative splice sites results in skipping of the alternative exon, whereas in panel E, PTB binding to sequences adjacent to constitutive exons tends to induce the inclusion of the alternative exon.

Table 1. Selected examples of regulation by hnRNP and hnRNP-like proteins

Mechanisms	hnRNP or hnRNP-Like	Target Genes	Antagonist	References
Negative Regulation				
Direct interference	• MBNL	• TNNT2	• U2AF2	219
	• Nova 1	• MAP4	• U1 snRNP	220
	• hnRNP A/B	• HIV-1	• U2AF65	221
	• hnRNP A1	• HIV-1	• U2 snRNP	222
	• Fox2	• CGRP	• U2AF65	223
	• CUG-BP2	• CFTR	• U2AF65	185
	• TDP-43	• CFTR	• U2AF65	183
	• HuR	• Fas	• U2AF65	70
	• Hu proteins	• NF-1	• U1 and U6	64
	• PTB	• β-tropomyosin	snRNP	264
	• hnRNP H	• NF-1	• U2AF65	224
	• hnRNP A1	• SMN2	• U1 snRNP	175
			• U2 snRNP	
Looping	• hnRNP A1	• hnRNP A1		38
	• hnRNP H			90
Nucleation	• hnRNP A1	• HIV-tat		57
	• PTB	• FGFR2		206
Blocking enhancer function	• hnRNP A1	• HIV-tat	• SRSF2	226
	• hnRNP A1	• β-tropomyosin	• SRSF1	227
	• hnRNP A1	• c-src	• SRSF1	228
	• hnRNP H	• FGFR2	• SRSF1	204
	• PTB	• MYPT1	• TIA-1	63
	• hnRNP L	• CD45	• SRSF1	88
	• hnRNP F/H	• α-tropomyosin	• SRSF7	229
	• hnRNP A1	• HIPK3	• Tra2β	230
	• hnRNP G	• TPM3	• Tra2β	231
Preventing intron or exon definition	• PTB, hnRNP A1, A2	• PKM		187
		• Calca		223
	• Fox1 and Fox2	• NF-1		64
	• Hu proteins	• c-src		67
	• PTB			
Positive Regulation				
Looping	• hnRNP A1			38
	• hnRNP H			37
Co-operative interaction	• hnRNP G	• SMN2	• Tra2β	80
	• Sam68	• CD44	• U2AF65	89
	• TIA1 and TIAR	• NF-1	• U1 and U6	64
			snRNP	
Blocking silencer function	• PTB	• hnRNP A1	• SRSF9	91

because the kinetic balance of splice site recognition is shifted in favor of the internal exons.[49] Thus, inhibiting splice site recognition of the regulated exon promotes skipping while a similar interference directed at splice sites of flanking constitutive exons can favor exon inclusion. A slightly different conclusion was reached in another recent study:[50] While repressed exons had PTB binding sites in or upstream of the regulated exon, included exons displayed PTB binding sites downstream, suggestive of enhancer function and reminiscent of the position-dependent effects of Fox2, Nova and MBNL proteins (see below). Although the reasons for this discrepancy is unclear, the little overlap in PTB targets between the two studies suggests that each study sampled distinct regulatory strategies.

Looping

Sequestering a splice site within a loop created by a secondary structure can strongly decrease splicing.[52] A loop can also be formed when proteins bound on either side of a splice site or an exon interact (Fig. 3C). hnRNP A1 and hnRNP H display this property and modulate 5′ splice site selection without significantly affecting the initial binding of U1 snRNP to competing sites.[37,52,53] A similar change in pre-mRNA conformation may be required for more efficient repression by PTB. PTB alone or in combination with other proteins may loop out one or both splice sites to repress exon inclusion (Fig. 3C). This would explain why PTB-mediated repression often requires binding sites upstream and downstream of (or in) alternative exons or splice sites. Moreover, insertion of a PTB site upstream of an exon is sufficient to convert a PTB-activated exon into a PTB-repressed exon.[50] Consistent with this model, FRET and NMR spectroscopy were used to show that one molecule of PTB can interact simultaneously with two distinct binding sites, and in this way loop out the intervening RNA (see Chapter 9 by Cléry and Allain for details).[54] Another group using a different pre-mRNA found that several PTB molecules were binding to distinct sites,[55] and suggested that additional hnRNP-like proteins such as Raver1 and MBNL may be needed to loop out the RNA portion located in between bound PTB proteins.[26,55,56]

Nucleation

Other occlusion models have been proposed. For hnRNP A1 and PTB for example, binding could trigger the adjacent binding of other proteins to create a repressive zone[57-59] (Fig. 3D). The binding map of A1 orthologues in *Drosophila* suggests that nucleation may occur in some cases of splicing regulation.[47] Likewise, while a recent RNA binding map for hnRNP C suggests that repression of splicing could arise from hnRNP C binding near a 3′ss, for other transcripts, skipped exons may be entirely covered by hnRNP C-containing particles.[51] When only the preceding intron is preferentially covered by hnRNP C, exon inclusion is favored.

Blocking Enhancer Function

Steric interference does not only occur at the splice sites. Many cases of exon skipping have associated hnRNP proteins binding close to, or overlapping with, exonic enhancers that are often binding sites for SR proteins (Fig. 3E) (Table 1). The hnRNP-like protein TIA1 and the related TIAR protein can stimulate 5′ss usage by binding to intronic U-rich sequences,[60] helping to recruit U1 snRNP by interacting with the U1C protein.[61,62] In the MYPT1 pre-mRNA, PTB can antagonize TIA1 binding and this competition is thought to determine tissue-specific splicing.[63] A similar antagonism between TIA1/TIAR and Hu proteins is occurring in the splicing control of neurofibromatosis exon 23a.[64]

Preventing Intron and Exon Definition

Negative regulation can also impact later steps of spliceosome assembly. In the c-src pre-mRNA, PTB represses the inclusion of alternative exon N1 by binding in the intron downstream.[65,66] This binding does not affect the interaction of U1 snRNP binding at the 5′ss but rather hinders association of U2AF at the downstream 3′ss, thereby interfering with intron definition (Fig. 3F).[67] In neuronal cells, repression is relieved by nPTB, the brain-specific paralogue of PTB.[68] When PTB binds to the regulated exon 6 in the Fas pre-mRNA, it can prevent exon inclusion by antagonizing exon definition (Fig. 3G).[61,69] A similar situation occurs with

HuR when its binding to an exonic silencer can antagonize a U1 snRNP-mediated improvement in U2AF binding across the exon.[70] In *Drosophila*, the hnRNP A1 orthologue hrp48 forms an exonic complex with PSI (P-element-specific inhibitor) and U1 snRNP to repress the authentic downstream 5'ss, possibly by sterically blocking the productive assembly of the spliceosome.[71] A similar mechanism implicating cryptic 5'ss may occur in the human Bcl-x pre-mRNA.[72] When Fox proteins bind upstream of a regulated exon they can inhibit splicing complex formation by preventing branchsite recognition by SF1, while their binding in the exon inhibits a later step of spliceosome assembly.[73] hnRNP L also provides a telling example in this category since, in collaboration with PSF, it can repress the inclusion of exon 4 in the CD45 pre-mRNA by favoring the assembly of a U1 and U2 snRNPs-containing complex and by preventing their use in later steps of spliceosome assembly (Fig. 3H).[74,75] Thus, the assembly of hnRNP-containing complexes in or near a regulated exon can interfere with splice site selection. Based on similar principles, inserting or positioning nonnatural binding sites for proteins in the vicinity of splice sites is an emerging technology to inhibit a specific splicing event and hence reprogram splicing decisions.[76,77]

Positive Regulation

Enhancer Function

As mentioned previously, the relationship between SR and hnRNP proteins is not always antagonistic. A co-operative interaction between RBMX and RBMY can stimulate exon inclusion in the SMN2 pre-mRNA most likely through an interaction with the SR protein Tra2β bound to a splicing enhancer.[78-80] Intronic binding sites for hnRNP F, H and L can also stimulate splicing,[81-87] but the mechanism by which stimulation occurs remains unclear. On the other hand, exonic binding sites for hnRNP L can also enhance the inclusion of an exon with weak splice sites, while repression occurs if the splice sites are strong.[88] Somewhat similarly, intronic binding sites for hnRNP H have been shown to stimulate splicing to an upsteam 5'ss more effectively if this site is of intermediate strength rather than weak or strong.[87] Sam68 can stimulate the recruitment of U2AF at a 3'ss.[89]

Looping

Although the looping out model mentioned above was invoked to explain the repression of splice sites, a loop also brings in closer proximity distantly located sequences. If these sequences contain splice sites, splicing between them could be stimulated[38] (Fig. 3C). Results consistent with this model have been obtained with individual binding sites for hnRNP A1 and hnRNP H,[37] as well as combinations of binding sites for these proteins.[90] The existence of heterotypic interactions, using RGG domains for example, considerably expands the possibilities for combinatorial modulation of splicing by different hnRNP and hnRNP-like proteins. Relevant to this model is the observation that many pre-mRNAs are controlled by several hnRNP A1 orthologues in *Drosophila*.[47] Since several of these splicing units harbor distantly spaced binding sites, heterotypic interactions between related proteins to control splice site selection may be a frequent occurrence.

Blocking Silencer Function

In one case, the usual alliance of SR proteins with enhancers and hnRNP proteins with repressors is completely reversed; in the hnRNP A1 pre-mRNA, an intron silencer element bound by SRp30c is antagonized by the binding of PTB, leading to re-activation of the 3'ss.[43,91]

RNA binding maps have been obtained for the hnRNP-like proteins Nova and Fox2. While the interaction of Nova and Fox2 upstream of a regulated exon correlates with exon skipping possibly because of interference with 3'ss usage, it is still unclear how the binding of Nova and Fox2 downstream of the regulated exon imposes exon inclusion. Similar enhancer effects have been reported for ESRP, MBNL and PTB.[50,92,93] Interestingly, the second RRM of PTB and the following interlinker domain appear important for this enhancing activity[50] and this RRM can interact with a specific region in the Raver1 protein.[56]

Linking hnRNP Protein Activity with ...

...Transcriptional Control

Functional coupling exists between transcription and splicing.[94-96] Since SR proteins have been implicated in this process,[97-100] it would not be too surprising if in some instances hnRNP and hnRNP-like proteins also relied on transcriptional coupling to enforce splicing regulation. Given the association of hnRNP C with the SWI/SNF chromatin remodelling complex,[101] and the role of other SWI/SNF components in alternative splicing,[102,103] hnRNP C may mediate some of its splicing effects cotranscriptionally. Other hnRNP proteins can bind promoters and associate with transcription factors to modulate transcription.[5-7,101,104-113] Thus, some hnRNP proteins could be co-ordinating splicing decisions cotranscriptionally either by affecting the speed of transcription, and hence modulating the timing with which competing splicing signals are emerging from the elongating RNA polymerase complex,[98,99] or they may be traveling with the RNA polymerase complex to be deposited preferentially at regulated exons.

...Signaling and Post-Pranslational Modifications

Another pathway that intersects with splicing control is signal transduction.[114-116] Activation of the MKK-$_{3/6}$-p38 stress-signaling pathway can lead to the hyperphosphorylation of hnRNP A1 and its cytoplasmic accumulation with an impact on alternative splicing.[117] Likewise, PTB can be phosphorylated at Ser-16 by protein kinase A, forcing PTB out of the nucleus,[118] and hnRNP K accumulates in the cytoplasm when phosphorylated by ERK.[119,120] The phosphorylation of Ser-52 in hnRNP L changes caspase-9 pre-mRNA splicing confering tumorigenic capacity to NSCLC cells.[121] The FAST kinase can phosphorylate TIA1/TIAR to stimulate the recruitment of U1 snRNP.[122] Sam68 phosphorylation modulates its activity in splicing control.[89,123,124] hnRNP proteins can also be sumoylated,[125-127] but the impact of this modification on splicing is unclear. Arginine methylation occurs on many RNA binding proteins including hnRNP A1, A2, K and U proteins.[128-132] The methyltransferase PRMT4 (a.k.a. CARM1) can methylate splicing and transcription elongation factors and affect splice site selection.[133,134] PRMT5 even links the circadian clock to the control of alternative splicing in plants.[135] However, a role for arginine methylation in the splicing function of hnRNP proteins remains for the moment hypothetical.

Since the specificity of action of splicing regulators likely relies on combinatorial strategies implicating interactions between different RNA binding proteins, post-translational modifications that change the localization and interaction properties of these proteins are likely to be of key importance when a cell must rapidly reprogram splicing decisions following a change in the environment that occurs during normal development or after stresses.

hnRNPs in Human Diseases

Since hnRNP and hnRNP-like proteins are crucial regulators of alternative pre-mRNA splicing, their relative stoichiometries in cells will ensure the proper expression of nearly all human proteins. Thus, even mild changes in the levels of these factors may lead to perturbations in alternative splicing patterns that may contribute to several human diseases. The following selected descriptions illustrate the crucial contributions of specific hnRNPs to diseases.

Myotonic Dystrophy

Myotonic dystrophy (DM) constitutes the most prevalent type of muscular atrophy in adults and is caused by an autosomal dominant mutation either in the 3' UTR of the DMPK gene (DM1) or in the first intron of the ZNF9 gene (DM2). The DMPK and ZNF9 mutations correspond respectively to CTG and CCTG repeat expansions, and become pathological when the number of repeats is superior to 38-50.[136] Transcripts with long CUG or CCUG tracts accumulate in nuclear foci. MBNL1 binds to and is sequestered by CUG repeats.[137] In contrast, CUG-BP1 is upregulated and stabilized in DM1 cells through a PKC-dependent pathway.[138] The depletion of MBNL and the overexpression of CUG-BP1 in mice reproduce many features of the DM1 phenotype.[139-141] The alternative splicing of many genes is altered in DM1 patients including the insulin receptor and the chloride channel genes which are respectively responsible for insulin resistance and myotonia.[142,143] Many of these splicing defects are antagonistically controlled by CELF (CUG-BP1 and CUG-BP2) and MBNL proteins.[144] For example, CELF proteins promote cTNT exon 5 inclusion and IR exon 11 skipping, whereas opposite effects are mediated by MBNL. Regulation of IR splicing occurs through distinct sequence elements, ruling out direct competition.[145,146] The co-ordinated regulation of splicing by MBNL and CUG-BP1 often requires other proteins like hnRNP H and Fox2.[144,147] Although several of the DM1-specific splicing changes have been observed in mice models, recent studies are suggesting an additional impact on transcription.[92,141,148]

Fragile X-Associated Tremor/Ataxia Syndrome

Fragile X-associated tremor/ataxia syndrome (FXTAS) is a neurodegenerative disorder caused by a CGG triplet repeat expansion in the FMR1 gene. Similarly to muscular dystrophy, the pathology would involve a toxic RNA gain-of-function in which mRNA with expanded repeats accumulate as inclusions in the nuclei of neurons and astrocytes. So far, various RNA binding proteins including hnRNP A2 and MBNL1 have been recovered from these inclusion bodies,[149] and two studies using a Drosophila model of FXTAS indicate that the binding of hnRNP A2 to CGG repeats recruits CUG-BP1.[150,151] In human cells however, it is Sam68 which is partially sequestered by these repeats leading to impaired Sam68-dependent alternative splicing.[152] Notably, the phosphorylation of Sam68 alters its RNA binding ability and disrupts its localization with CGG repeats.

Schizophrenia

Schizophrenia is a complex chronic mental illness. Aberrant alternative splicing has been observed in genes involved in neurogenesis and neural functions.[153] Because myelination defects contribute to the development of schizophrenia and QKI is a splicing regulator of myelin compo-nents,[154] the diminished expression, as well as perturbations in the alternative splicing of the QKI transcript per se in patients makes QKI a prime candidate for promoting schizophrenia-specific splicing defects.[155] Interestingly, QKI also directly controls the translation of hnRNP A1, which is implicated in the splicing control of the myelin-associated glycoprotein (MAG) pre-mRNA.[156]

Frontotemporal Lobar Degeneration

Although frontotemporal lobar degeneration (FTLD) is often caused by mutations in the MAPT/tau gene that can affect its splicing, another form of FTLD (FTLD-TDP) is associated with mutations, improper localization and overmodification (phosphorylation and ubiquitina-tion) of TDP-43.[157] Likewise, mutations in TDP-43 and FUS/TLS and aberrant cytoplasmic aggregation in neural and glial cells have been uncovered in cases of amyotrophic lateral sclerosis (ALS) also known as Lou Gehrig's disease.[158-160] Since the hnRNP-like protein TDP-43 can impact splicing decisions, these neuronal pathologies could result at least in part from splicing defects due to the altered function of TDP-43. However, it remains to be shown if the abnormal splicing events occurring in FTLD and ALS[161,162] are caused by aberrant TDP-43 function (see Chapter 3 by Rogelj et al for more details on TDP-43 function).

Spinal Muscular Atrophy and Cystic Fibrosis

The following two diseases (spinal muscular atrophy and cystic fibrosis) are not caused by the aberrant expression, sequestration or mutations of hnRNP or hnRNP-like proteins. However, the alternative splicing events that contribute to the diseases are controlled by these proteins. A better undertanding of the splicing regulation of these genes therefore offers opportunities for interventions aimed at correcting splicing defects.

Spinal Muscular Atrophy

Spinal muscular atrophy (SMA) is characterized by the degeneration of motor neurons and is caused by a loss of the SMN1 gene that creates a deficiency in the SMN protein.[163] This deficiency alters the relative abundance of multiple snRNAs and causes widespread defects in alternative splicing.[164] The paralogous SMN2 gene contains a nucleotide difference in exon 7 that promotes frequent skipping, leading to the production of an inactive protein, and not enough correct SMN protein to rescue the deficiency created by the loss of SMN1 (Fig. 5A). The C to U change in exon 7 has been proposed to inactivate an enhancer bound by SRSF1 or to create a silencer bound by hnRNP A1.[165,166] Other positive regulators have been uncovered including Tra2β, SRSF9, hnRNP G, hnRNP Q1, TDP-43 and TIA1.[79,80,167-170] Sam68 was also identified as a regulator acting through a novel silencer that would collaborate with the hnRNP A1 binding element.[171] hnRNP A1 was also found to confer repression through a second binding site in the downstream intron.[172] An interaction between these two distinctly bound A1 molecules may be important for more efficient repression of exon 7 inclusion.[173-175] Preventing A1 binding at this site with an antisense oligonucleotide has proven effective in stimulating the inclusion of SMN2 exon 7,[176-178] and improvement of phenotypes are observed after the systemic or targeted administration of the oligonucleotide in model mice.[176,179]

Cystic Fibrosis

Cystic fibrosis (CF) is caused by mutations in the CFTR gene that encodes a chloride channel required for proper secretory epithelial function of the lung. In some patients, elevated exon 9 skipping produces a nonfunctional protein whose levels correlate with the severity of the disease.[180] While TIA1 can favor exon 9 inclusion, other splicing factors including TDP-43, hnRNP A1, PTB, SR and CUG-BP proteins can promote exon 9 skipping, some by binding to the polymorphic polypyrimidine tract at the 3′ss.[181-185] Hopefully, preventing one or several of these interactions to mimic what is being accomplished with SMN2 may stimulate the inclusion of exon 9 and improve CFTR function in deficient cells and in patients.

Cancer

Cancer is without doubts the disease that is receiving most of the attention in terms of altered hnRNP expression and aberrant splicing profiling. Because of the complexity of the disease and its many different types, the information remains however largely incomplete. Changes in alternative splicing have consistently been associated with cancer,[186] and shifts in the proportion of splice variants have been described for genes implicated in all aspects of tumor biology including cell cycle control, apoptosis, metabolism, cell migration, metastasis, and angiogenesis.[187,188] While some of these changes may be caused by mutations at splice sites and splicing regulatory signals, many studies have reported significant alterations in the expression of hnRNP and hnRNP-like proteins in cancer tissues and during tumor development.[189-193] Large-scale perturbations of splicing in ovarian and breast cancers were recently associated with the downregulation of Fox2,[194] correlating with specific breast cancer subtypes.[195] Although these splicing changes are useful as biomarkers, the biggest challenge is now to determine the contribution of these splicing disturbances to the disease. This knowledge will reap high benefits because it will increase the number of cancer targets and should stimulate the development of novel approaches to correct splicing defects.While a link between hnRNP protein expression and cell proliferation was established decades ago,[196] a clear contribution of hnRNP proteins in the splicing regulation of cancer-relevant genes was obtained only recently. The following examples provide possibly the best causative link between the expression of hnRNPs or hnRNP-like proteins and cancer.

Pyruvate Kinase M

The splicing decision that switches the production of pyruvate kinase M1 (PKM1) to the embryonic PKM2 form and promotes aerobic glycolysis is vital for tumor cell growth.[197] This switch occurs by repressing the use of exon 9, which leads to the inclusion of the mutually exclusive exon 10 neighbor. The binding of PTB, hnRNP A1 and hnRNP A2 to sequences flanking exon 9 represses its inclusion to stimulate the production of PKM2[198,199] (Fig. 5). This switch is enforced by c-Myc, which activates the transcription of A1, A2 and PTB and may be responsible for the elevated levels of these RNA binding proteins in many cancers.[198,200,201]

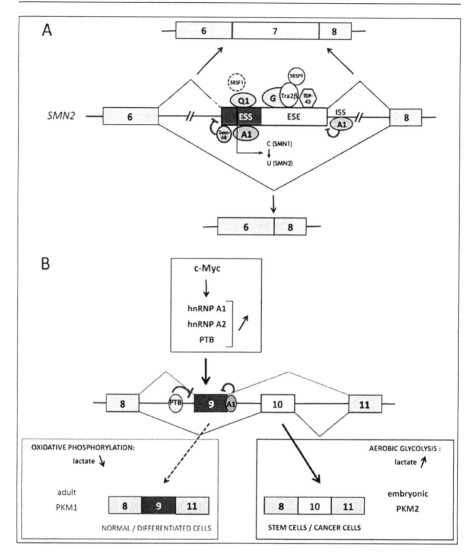

Figure 5. Examples of splicing regulation relevant to diseases. A) Regulators of SMN2 alternative splicing. Relative to the SMN1 gene, a C to U single nucleotide substitution at position 6 in SMN2 exon 7 induces exon skipping, leading to a functional deficiency of SMN proteins. The C to U mutation has been suggested to disrupt an enhancer that binds SRSF1 in SMN1 and creates an exonic splicing enhancer (ESS) bound by hnRNP A1. hnRNP A1 also binds to an intronic splicing silencer (ISS) located just downstream of exon 7. Interactions between bound A1 molecules could enhance repression. Sam68 binds just upstream of hnRNP A1 to repress exon 7 inclusion. SMN2 exon 7 skipping can be antagonized by overexpression of Tra2β and hnRNP G, the latter exerting its activity through specific interaction with Tra2β. SRSF9 and TDP-43 also favor exon 7 inclusion. B) Alternative splicing regulation of pyruvate kinase M (PKM). The PKM gene comprises two mutually exclusive exons whose alternative splicing is differentially regulated between normal (exon 9 included) and cancer cells (exon 10 included). In tumor cells, hnRNP A1, A2 and PTB bind to intronic regions flanking exon 9 and drive an embryonic splicing switch by repressing exon 9 inclusion, replacing the adult PKM1 isoform by the PKM2 splice variant. The c-Myc oncogene upregulates transcription of hnRNP A1, A2 and PTB, raising the PKM2/PKM1 ratio. This PKM1 to PKM2 switch stimulates aerobic glycolysis, which enhances tumor cell proliferation.

Epithelial-to-Mesenchymal Transition

ESRP1 and ESRP2 are two hnRNP-like RNA binding proteins that play a critical role in enforcing the epithelial-to-mesenchymal transition (EMT) that occurs during the metastatic process.[93] Indeed, ESRP1 and ESRP2 promote the epithelial switch towards FGFR2 exon IIIb inclusion, as well as a battery of other epithelial-specific alternative splicing events.[202,203] hnRNP F/H, K and Fox2 also act to silence the mesenchymal-specific FGFR2 exon IIIc,[204] whereas hnRNP A1 and PTB have been implicated in the silencing of exon IIIb.[205,206] While ESRP binding sites downstream of a regulated exon are stimulatory, the most prevalent regulatory strategy is to have ESRP binding sites in the regulated exon to favor skipping.[93] Interestingly, the Fox/ESRP collaboration appears to extend beyond the regulation of FGFR2 since Fox binding motifs are often associated with ESRP regulated exons, suggesting that this combinatorial strategies is an important aspect of of the splicing regulation during EMT.

Oncogenes

The extended family of hnRNP proteins also participates in the alternative splicing regulation of oncogenes. Sam68 enhances the production of the more oncogenic cyclin D1b splice variant by binding to the proximal region of intron 4, and its upregulation is associated with increased D1b expression in human prostate cancers.[207] Moreover, Sam68 controls the alternative splicing of SRSF1 to produce a variant mRNA susceptible to degradation by the nonsense RNA decay pathway.[208] Since SRSF1 controls the splicing of the macrophage-stimulating receptor Ron to produce an isoform that confers an invasive phenotype,[209] the higher levels of Sam68 reported in breast, renal and prostate cancers[210] may impact SRSF1 levels and in turn affect the alternative splicing of Ron. Sam68 is therefore becoming a recurrent player in cancer-relevant splicing; it was first involved along with hnRNP A1 in regulating the production of isoforms of the cell-cell adhesion glycoprotein CD44 whose expression is associated with the metastatic potential of cancer cells.[124,211]

Chemoresistance

Intriguingly, the splicing of thymidine phosphorylase is inhibited by hnRNP H proteins.[212] Since this enzyme is essential to convert an intermediary metabolite of capecitabine to 5-fluorouracil, the expression of hnRNP H into drug sensitive cell lines promotes drug resistance. hnRNP A2 modulates the epithelial-mesenchymal transition in lung cancer cell lines and its expression is downregulated in chemoresistant ovarian cancer tissues.[213,214] All anticancer drugs tested so far affect the alternative splicing of many apoptotic regulators including Bcl-x,[215] sometimes by activating the p53-dependent DNA damage response pathway.[216] Alterations in the expression of modifications of several hnRNP (A1, F, H, K) and hnRNP-like (Sam68) proteins that regulate the splicing of Bcl-x[85,123,217] and other apoptotic regulators[121,218] offer a multitude of potential routes for a cancer cell to achieve drug resistance.

Conclusion

Our understanding of the function of hnRNP and hnRNP-related proteins has progressed rapidly in the last decade. The initial involvement of hnRNP proteins in alternative splicing control has now been abundantly confirmed although the evidence to support their role in generic splicing has remained sporadic. Hopefully, the continued application of high-throughput analyses will accelerate the identification of pre-mRNA targets and the production of maps illustrating the extensive interaction of these proteins with the transcriptome as well as the dynamic changes in binding occurring with time and in a variety of conditions. Expanding the number of examples where hnRNP proteins affect splicing decisions will also contribute to understanding their various mechanisms of action. Exploring the role of hnRNP proteins in coupling transcription with splicing decisions and assessing the impact of post-translational modifications on their function will be essential to obtain a full view of the complex molecular pathways and regulatory networks to which these proteins partake. Future challenges will be to determine how these intricate levels of regulation are orchestrated during normal animal development and to understand the contribution of the different hnRNP and hnRNP-like proteins to an increasing variety of pathological conditions.

References

1. Beyer AL, Christensen ME, Walker BW et al. Identification and characterization of the packaging proteins of core 40S hnRNP particles. Cell 1977; 11(1):127-138.
2. Pinol-Roma S, Choi YD, Matunis MJ et al. Immunopurification of heterogeneous nuclear ribonucleoprotein particles reveals an assortment of RNA-binding proteins. Genes Dev 1988; 2(2):215-227.
3. Dreyfuss G, Matunis MJ, Pinol-Roma S et al. hnRNP proteins and the biogenesis of mRNA. Annu Rev Biochem 1993; 62:289-321.
4. Bonnal S, Pileur F, Orsini C et al. Heterogeneous nuclear ribonucleoprotein A1 is a novel internal ribosome entry site trans-acting factor that modulates alternative initiation of translation of the fibroblast growth factor 2 mRNA. J Biol Chem 2005; 280(6):4144-4153.
5. Campillos M, Lamas JR, Garcia MA et al. Specific interaction of heterogeneous nuclear ribonucleoprotein A1 with the -219T allelic form modulates APOE promoter activity. Nucleic Acids Res 2003; 31(12):3063-3070.
6. Lau JS, Baumeister P, Kim E et al. Heterogeneous nuclear ribonucleoproteins as regulators of gene expression through interactions with the human thymidine kinase promoter. J Cell Biochem 2000; 79(3):395-406.
7. Xia H. Regulation of gamma-fibrinogen chain expression by heterogeneous nuclear ribonucleoprotein A1. J Biol Chem 2005; 280(13):13171-13178.
8. LaBranche H, Dupuis S, Ben-David Y et al. Telomere elongation by hnRNP A1 and a derivative that interacts with telomeric repeats and telomerase. Nat Genet 1998; 19(2):199-202.
9. Zhang QS, Manche L, Xu RM et al. hnRNP A1 associates with telomere ends and stimulates telomerase activity. RNA 2006; 12(6):1116-1128.
10. Nakielny S, Dreyfuss G. The hnRNP C proteins contain a nuclear retention sequence that can override nuclear export signals. J Cell Biol 1996; 134(6):1365-1373.
11. Rajagopalan LE, Malter JS. Turnover and translation of in vitro synthesized messenger RNAs in transfected, normal cells. J Biol Chem 1996; 271(33):19871-19876.
12. Hamilton BJ, Nichols RC, Tsukamoto H et al. hnRNP A2 and hnRNP L bind the 3'UTR of glucose transporter 1 mRNA and exist as a complex in vivo. Biochem Biophys Res Commun 1999; 261(3):646-651.
13. Pinol-Roma S, Dreyfuss G. Shuttling of pre-mRNA binding proteins between nucleus and cytoplasm. Nature 1992; 355(6362):730-732.
14. Li HP, Zhang X, Duncan R et al. Heterogeneous nuclear ribonucleoprotein A1 binds to the transcription-regulatory region of mouse hepatitis virus RNA. Proc Natl Acad Sci U S A 1997; 94(18):9544-9549.
15. Hamilton BJ, Burns CM, Nichols RC et al. Modulation of AUUUA response element binding by heterogeneous nuclear ribonucleoprotein A1 in human T-lymphocytes. The roles of cytoplasmic location, transcription and phosphorylation. J Biol Chem 1997; 272(45):28732-28741.
16. Honore B, Rasmussen HH, Vorum H et al. Heterogeneous nuclear ribonucleoproteins H, H', and F are members of a ubiquitously expressed subfamily of related but distinct proteins encoded by genes mapping to different chromosomes. J Biol Chem 1995; 270(48):28780-28789.
17. Perez I, McAfee JG, Patton JG. Multiple RRMs contribute to RNA binding specificity and affinity for polypyrimidine tract binding protein. Biochemistry 1997; 36(39):11881-11890.
18. Gorlach M, Wittekind M, Beckman RA et al. Interaction of the RNA-binding domain of the hnRNP C proteins with RNA. EMBO J 1992; 11(9):3289-3295.
19. Wan L, Kim JK, Pollard VW et al. Mutational definition of RNA-binding and protein-protein interaction domains of heterogeneous nuclear RNP C1. J Biol Chem 2001; 276(10):7681-7688.
20. McAfee JG, Shahied-Milam L, Soltaninassab SR et al. A major determinant of hnRNP C protein binding to RNA is a novel bZIP-like RNA binding domain. RNA 1996; 2(11):1139-1152.
21. Shahied-Milam L, Soltaninassab SR, Iyer GV et al. The heterogeneous nuclear ribonucleoprotein C protein tetramer binds U1, U2 and U6 snRNAs through its high affinity RNA binding domain (the bZIP-like motif). J Biol Chem 1998; 273(33):21359-21367.
22. Kiledjian M, Dreyfuss G. Primary structure and binding activity of the hnRNP U protein: binding RNA through RGG box. EMBO J 1992; 11(7):2655-2664.
23. Iko Y, Kodama TS, Kasai N et al. Domain architectures and characterization of an RNA-binding protein, TLS. J Biol Chem 2004; 279(43):44834-44840.
24. Teplova M, Patel DJ. Structural insights into RNA recognition by the alternative-splicing regulator muscleblind-like MBNL1. Nat Struct Mol Biol 2008; 15(12):1343-1351.
25. Kipp M, Schwab BL, Przybylski M et al. Apoptotic cleavage of scaffold attachment factor A (SAF-A) by caspase-3 occurs at a noncanonical cleavage site. J Biol Chem 2000; 275(7):5031-5036.
26. Gromak N, Rideau A, Southby J et al. The PTB interacting protein raver1 regulates alpha-tropomyosin alternative splicing. EMBO J 2003; 22(23):6356-6364.

27. Bang ML, Centner T, Fornoff F et al. The complete gene sequence of titin, expression of an unusual approximately 700-kDa titin isoform, and its interaction with obscurin identify a novel Z-line to I-band linking system. Circ Res 2001; 89(11):1065-1072.
28. Lander ES, Linton LM, Birren B et al. Initial sequencing and analysis of the human genome. Nature 2001; 409(6822):860-921.
29. Jurica MS, Moore MJ. Pre-mRNA splicing: awash in a sea of proteins. Mol Cell 2003; 12(1):5-14.
30. Bessonov S, Anokhina M, Will CL et al. Isolation of an active step I spliceosome and composition of its RNP core. Nature 2008; 452(7189):846-850.
31. Sierakowska H, Szer W, Furdon PJ et al. Antibodies to hnRNP core proteins inhibit in vitro splicing of human beta-globin pre-mRNA. Nucleic Acids Res 1986; 14(13):5241-5254.
32. Gamberi C, Izaurralde E, Beisel C et al. Interaction between the human nuclear cap-binding protein complex and hnRNP F. Mol Cell Biol 1997; 17(5):2587-2597.
33. Choi YD, Grabowski PJ, Sharp PA et al. Heterogeneous nuclear ribonucleoproteins: role in RNA splicing. Science 1986; 231(4745):1534-1539.
34. Mourelatos Z, Abel L, Yong J et al. SMN interacts with a novel family of hnRNP and spliceosomal proteins. EMBO J 2001; 20(19):5443-5452.
35. Michaud S, Reed R. An ATP-independent complex commits pre-mRNA to the mammalian spliceosome assembly pathway. Genes Dev 1991; 5(12B):2534-2546.
36. Bennett M, Pinol-Roma S, Staknis D et al. Differential binding of heterogeneous nuclear ribonucleoproteins to mRNA precursors prior to spliceosome assembly in vitro. Mol Cell Biol 1992; 12(7):3165-3175.
37. Martinez-Contreras R, Fisette JF, Nasim FU et al. Intronic binding sites for hnRNP A/B and hnRNP F/H proteins stimulate pre-mRNA splicing. PLoS Biol 2006; 4(2):e21.
38. Blanchette M, Chabot B. Modulation of exon skipping by high-affinity hnRNP A1-binding sites and by intron elements that repress splice site utilization. EMBO J 1999; 18(7):1939-1952.
39. Wang ET, Sandberg R, Luo S et al. Alternative isoform regulation in human tissue transcriptomes. Nature 2008; 456(7221):470-476.
40. Pan Q, Shai O, Lee LJ et al. Deep surveying of alternative splicing complexity in the human transcriptome by high-throughput sequencing. Nat Genet 2008; 40(12):1413-1415.
41. Black DL. Mechanisms of alternative pre-messenger RNA splicing. Annual Rev Biochem 2003; 72:291-336.
42. Lin S, Fu XD. SR proteins and related factors in alternative splicing. Adv Exp Med Biol 2007; 623:107-122.
43. Simard MJ, Chabot B. SRp30c is a repressor of 3' splice site utilization. Mol Cell Biol 2002; 22(12):4001-4010.
44. Kanopka A, Mühlemann O, Aküsjarvi G. Inhibition by SR proteins of splicing of a regulated adenovirus pre-mRNA. Nature 1996; 381(6582):535-538.
45. Shin C, Manley JL. The SR protein SRp38 represses splicing in M phase cells. Cell 2002; 111(3):407-417.
46. Blanchette M, Green RE, Brenner SE et al. Global analysis of positive and negative pre-mRNA splicing regulators in Drosophila. Genes Dev 2005; 19(11):1306-1314.
47. Blanchette M, Green RE, MacArthur S et al. Genome-wide analysis of alternative pre-mRNA splicing and RNA-binding specificities of the Drosophila hnRNP A/B family members. Mol Cell 2009; 33(4):438-449.
48. Venables JP, Koh CS, Froehlich U et al. Multiple and specific mRNA processing targets for the major human hnRNP proteins. Mol Cell Biol 2008; 28(19):6033-6043.
49. Xue Y, Zhou Y, Wu T et al. Genome-wide analysis of PTB-RNA interactions reveals a strategy used by the general splicing repressor to modulate exon inclusion or skipping. Mol Cell 2009; 36(6):996-1006.
50. Llorian M, Schwartz S, Clark TA et al. Position-dependent alternative splicing activity revealed by global profiling of alternative splicing events regulated by PTB. Nat Struct Mol Biol 2010; 17(9):1114-1123.
51. Konig J, Zarnack K, Rot G et al. iCLIP reveals the function of hnRNP particles in splicing at individual nucleotide resolution. Nat Struct Mol Biol 2010; 17(7):909-915.
52. Nasim FU, Hutchison S, Cordeau M et al. High-affinity hnRNP A1 binding sites and duplex-forming inverted repeats have similar effects on 5' splice site selection in support of a common looping out and repression mechanism. RNA 2002; 8(8):1078-1089.
53. Chabot B, Blanchette M, Lapierre I et al. An intron element modulating 5' splice site selection in the hnRNP A1 pre-mRNA interacts with hnRNP A1. Mol Cell Biol 1997; 17(4):1776-1786.
54. Lamichhane R, Daubner GM, Thomas-Crusells J et al. RNA looping by PTB: evidence using FRET and NMR spectroscopy for a role in splicing repression. Proc Natl Acad Sci U S A 2010; 107(9):4105-4110.
55. Cherny D, Gooding C, Eperon GE et al. Stoichiometry of a regulatory splicing complex revealed by single-molecule analyses. EMBO J 2010; 29(13):2161-2172.
56. Rideau AP, Gooding C, Simpson PJ et al. A peptide motif in Raver1 mediates splicing repression by interaction with the PTB RRM2 domain. Nat Struct Mol Biol 2006; 13(9):839-848.

57. Zhu J, Mayeda A, Krainer AR. Exon identity established through differential antagonism between exonic splicing silencer-bound hnRNP A1 and enhancer-bound SR proteins. Mol Cell 2001; 8(6):1351-1361.

58. Wagner EJ, Garcia-Blanco MA. Polypyrimidine tract binding protein antagonizes exon definition. Mol Cell Biol 2001; 21(10):3281-3288.

59. Okunola HL, Krainer AR. Cooperative-binding and splicing-repressive properties of hnRNP A1. Mol Cell Biol 2009; 29(20):5620-5631.

60. Wang Z, Kayikci M, Briese M et al. iCLIP predicts the dual splicing effects of TIA-RNA interactions. PLoS Biol 2010; 8(10):e1000530.

61. Izquierdo JM, Majos N, Bonnal S et al. Regulation of Fas alternative splicing by antagonistic effects of TIA-1 and PTB on exon definition. Mol Cell 2005; 19(4):475-484.

62. Forch P, Puig O, Martinez C et al. The splicing regulator TIA-1 interacts with U1-C to promote U1 snRNP recruitment to 5′ splice sites. EMBO J 2002; 21(24):6882-6892.

63. Shukla S, Del Gatto-Konczak F, Breathnach R et al. Competition of PTB with TIA proteins for binding to a U-rich cis-element determines tissue-specific splicing of the myosin phosphatase targeting subunit 1. RNA 2005; 11(11):1725-1736.

64. Zhu H, Hinman MN, Hasman RA et al. Regulation of neuron-specific alternative splicing of neurofibromatosis type 1 pre-mRNA. Mol Cell Biol 2008; 28(4):1240-1251.

65. Chou MY, Underwood JG, Nikolic J et al. Multisite RNA binding and release of polypyrimidine tract binding protein during the regulation of c-src neural-specific splicing. Mol Cell 2000; 5(6):949-957.

66. Chan RC, Black DL. The polypyrimidine tract binding protein binds upstream of neural cell-specific c-src exon N1 to repress the splicing of the intron downstream. Mol Cell Biol 1997; 17(8):4667-4676.

67. Sharma S, Falick AM, Black DL. Polypyrimidine tract binding protein blocks the 5′ splice site-dependent assembly of U2AF and the prespliceosomal E complex. Mol Cell 2005; 19(4):485-496.

68. Markovtsov V, Nikolic JM, Goldman JA et al. Cooperative assembly of an hnRNP complex induced by a tissue-specific homolog of polypyrimidine tract binding protein. Mol Cell Biol 2000; 20(20):7463-7479.

69. Spellman R, Smith CW. Novel modes of splicing repression by PTB. Trends Biochem Sci 2006; 31(2):73-76.

70. Izquierdo JM. Hu antigen R (HuR) functions as an alternative pre-mRNA splicing regulator of Fas apoptosis-promoting receptor on exon definition. J Biol Chem 2008; 283(27):19077-19084.

71. Siebel CW, Admon A, Rio DC. Soma-specific expression and cloning of PSI, a negative regulator of P element pre-mRNA splicing. Genes Dev 1995; 9(3):269-283.

72. Cloutier P, Toutant J, Shkreta L et al. Antagonistic effects of the SRp30c protein and cryptic 5′ splice sites on the alternative splicing of the apoptotic regulator Bcl-x. J Biol Chem 2008; 283(31):21315-21324.

73. Zhou HL, Lou H. Repression of prespliceosome complex formation at two distinct steps by Fox-1/Fox-2 proteins. Mol Cell Biol 2008; 28(17):5507-5516.

74. Melton AA, Jackson J, Wang J et al. Combinatorial control of signal-induced exon repression by hnRNP L and PSF. Mol Cell Biol 2007; 27(19):6972-6984.

75. House AE, Lynch KW. An exonic splicing silencer represses spliceosome assembly after ATP-dependent exon recognition. Nat Struct Mol Biol 2006; 13(10):937-944.

76. Villemaire J, Dion I, Elela SA et al. Reprogramming alternative pre-messenger RNA splicing through the use of protein-binding antisense oligonucleotides. J Biol Chem 2003; 278(50):50031-50039.

77. Culler SJ, Hoff KG, Smolke CD. Reprogramming cellular behavior with RNA controllers responsive to endogenous proteins. Science 2010; 330(6008):1251-1255.

78. Venables JP, Elliott DJ, Makarova OV et al. RBMY, a probable human spermatogenesis factor, and other hnRNP G proteins interact with Tra2beta and affect splicing. Hum Mol Genet 2000; 9(5):685-694.

79. Hofmann Y, Lorson CL, Stamm S et al. Htra2-beta 1 stimulates an exonic splicing enhancer and can restore full-length SMN expression to survival motor neuron 2 (SMN2). Proc Natl Acad Sci U S A 2000; 97(17):9618-9623.

80. Hofmann Y, Wirth B. hnRNP-G promotes exon 7 inclusion of survival motor neuron (SMN) via direct interaction with Htra2-beta1. Hum Mol Genet 2002; 11(17):2037-2049.

81. Chou MY, Rooke N, Turck CW et al. hnRNP H is a component of a splicing enhancer complex that activates a c-src alternative exon in neuronal cells. Mol Cell Biol 1999; 19(1):69-77.

82. Min H, Chan RC, Black DL. The generally expressed hnRNP F is involved in a neural-specific pre-mRNA splicing event. Genes Dev 1995; 9(21):2659-2671.

83. Hui J, Stangl K, Lane WS et al. HnRNP L stimulates splicing of the eNOS gene by binding to variable-length CA repeats. Nat Struct Biol 2003; 10(1):33-37.

84. Cheli Y, Kunicki TJ. hnRNP L regulates differences in expression of mouse integrin alpha2beta1. Blood 2006; 107(11):4391-4398.

85. Garneau D, Revil T, Fisette JF et al. Heterogeneous nuclear ribonucleoprotein F/H proteins modulate the alternative splicing of the apoptotic mediator Bcl-x. J Biol Chem 2005; 280(24):22641-22650.

86. Hastings ML, Wilson CM, Munroe SH. A purine-rich intronic element enhances alternative splicing of thyroid hormone receptor mRNA. RNA 2001; 7(6):859-874.

87. Xiao X, Wang Z, Jang M et al. Splice site strength-dependent activity and genetic buffering by poly-G runs. Nat Struct Mol Biol 2009; 16(10):1094-1100.

88. Motta-Mena LB, Heyd F, Lynch KW. Context-dependent regulatory mechanism of the splicing factor hnRNP L. Mol Cell 2010; 37(2):223-234.

89. Tisserant A, Konig H. Signal-regulated Pre-mRNA occupancy by the general splicing factor U2AF. PLoS One 2008; 3(1):e1418.

90. Fisette JF, Toutant J, Dugre-Brisson S et al. hnRNP A1 and hnRNP H can collaborate to modulate 5' splice site selection. RNA 2010; 16(1):228-238.

91. Paradis C, Cloutier P, Shkreta L et al. hnRNP I/PTB can antagonize the splicing repressor activity of SRp30c. RNA 2007; 13:1287-1300.

92. Du H, Cline MS, Osborne RJ et al. Aberrant alternative splicing and extracellular matrix gene expression in mouse models of myotonic dystrophy. Nat Struct Mol Biol 2010; 17(2):187-193.

93. Warzecha CC, Jiang P, Amirikian K et al. An ESRP-regulated splicing programme is abrogated during the epithelial-mesenchymal transition. EMBO J 2010; 29(19):3286-3300.

94. Kornblihtt AR. Promoter usage and alternative splicing. Curr Opin Cell Biol 2005; 17(3):262-268.

95. Listerman I, Sapra AK, Neugebauer KM. Cotranscriptional coupling of splicing factor recruitment and precursor messenger RNA splicing in mammalian cells. Nat Struct Mol Biol 2006; 13(9):815-822.

96. Bentley DL. Rules of engagement: co-transcriptional recruitment of pre-mRNA processing factors. Curr Opin Cell Biol 2005; 17(3):251-256.

97. Cramer P, Caceres JF, Cazalla D et al. Coupling of transcription with alternative splicing: RNA pol II promoters modulate SF2/ASF and 9G8 effects on an exonic splicing enhancer. Mol Cell 1999; 4(2):251-258.

98. Nogues G, Kadener S, Cramer P et al. Transcriptional activators differ in their abilities to control alternative splicing. J Biol Chem 2002; 277(45):43110-43114.

99. de la Mata M, Alonso CR, Kadener S et al. A slow RNA polymerase II affects alternative splicing in vivo. Mol Cell 2003; 12(2):525-532.

100. Das R, Yu J, Zhang Z et al. SR proteins function in coupling RNAP II transcription to pre-mRNA splicing. Mol Cell 2007; 26(6):867-881.

101. Mahajan MC, Narlikar GJ, Boyapaty G et al. Heterogeneous nuclear ribonucleoprotein C1/C2, MeCP1, and SWI/SNF form a chromatin remodeling complex at the beta-globin locus control region. Proc Natl Acad Sci U S A 2005; 102(42):15012-15017.

102. Batsche E, Yaniv M, Muchardt C. The human SWI/SNF subunit Brm is a regulator of alternative splicing. Nat Struct Mol Biol 2006; 13(1):22-29.

103. Auboeuf D, Batsche E, Duterre M et al. Coregulators: transducing signal from transcription to alternative splicing. Trends Endocrinol Metab 2007; 18(3):122-129.

104. Wei CC, Zhang SL, Chen YW et al. Heterogeneous nuclear ribonucleoprotein k modulates angiotensinogen gene expression in kidney cells. J Biol Chem 2006; 281(35):25344-25355.

105. Moumen A, Masterson P, O'Connor MJ et al. hnRNP K: an HDM2 target and transcriptional coactivator of p53 in response to DNA damage. Cell 2005; 123(6):1065-1078.

106. Yoshida T, Makino Y, Tamura T. Association of the rat heterogeneous nuclear RNA-ribonucleoprotein F with TATA-binding protein. FEBS Lett 1999; 457(2):251-254.

107. Mattern KA, van Goethem RE, de Jong L et al. Major internal nuclear matrix proteins are common to different human cell types. J Cell Biochem 1997; 65(1):42-52.

108. Spraggon L, Dudnakova T, Slight J et al. hnRNP-U directly interacts with WT1 and modulates WT1 transcriptional activation. Oncogene 2006.

109. Kukalev A, Nord Y, Palmberg C et al. Actin and hnRNP U cooperate for productive transcription by RNA polymerase II. Nat Struct Mol Biol 2005; 12(3):238-244.

110. Kim MK, Nikodem VM. hnRNP U inhibits carboxy-terminal domain phosphorylation by TFIIH and represses RNA polymerase II elongation. Mol Cell Biol 1999; 19(10):6833-6844.

111. Swinburne IA, Meyer CA, Liu XS et al. Genomic localization of RNA binding proteins reveals links between pre-mRNA processing and transcription. Genome Res 2006; 16(7):912-921.

112. Law WJ, Cann KL, Hicks GG. TLS, EWS and TAF15: a model for transcriptional integration of gene expression. Brief Funct Genomic Proteomic 2006; 5(1):8-14.

113. Uranishi H, Tetsuka T, Yamashita M et al. Involvement of the pro-oncoprotein TLS (translocated in liposarcoma) in nuclear factor-kappa B p65-mediated transcription as a coactivator. J Biol Chem 2001; 276(16):13395-13401.

114. Shin C, Manley JL. Cell signalling and the control of pre-mRNA splicing. Nat Rev Mol Cell Biol 2004; 5(9):727-738.

115. Stamm S. Regulation of alternative splicing by reversible protein phosphorylation. J Biol Chem 2008; 283(3):1223-1227.
116. Blaustein M, Pelisch F, Srebrow A. Signals, pathways and splicing regulation. Int J Biochem Cell Biol 2007; 39(11):2031-2048.
117. van der Houven van Oordt W, Diaz-Meco MT, Lozano J et al. The MKK(3/6)-p38-signaling cascade alters the subcellular distribution of hnRNP A1 and modulates alternative splicing regulation. J Cell Biol 2000; 149(2):307-316.
118. Xie J, Lee JA, Kress TL et al. Protein kinase A phosphorylation modulates transport of the polypyrimidine tract-binding protein. Proc Natl Acad Sci U S A 2003; 100(15):8776-8781.
119. Mikula M, Karczmarski J, Dzwonek A et al. Casein kinases phosphorylate multiple residues spanning the entire hnRNP K length. Biochim Biophys Acta 2006; 1764(2):299-306.
120. Habelhah H, Shah K, Huang L et al. ERK phosphorylation drives cytoplasmic accumulation of hnRNP-K and inhibition of mRNA translation. Nat Cell Biol 2001; 3(3):325-330.
121. Goehe RW, Shultz JC, Murudkar C et al. hnRNP L regulates the tumorigenic capacity of lung cancer xenografts in mice via caspase-9 pre-mRNA processing. J Clin Invest 2010; 120(11):3923-3939.
122. Izquierdo JM, Valcarcel J. Fas-activated serine/threonine kinase (FAST K) synergizes with TIA-1/TIAR proteins to regulate Fas alternative splicing. J Biol Chem 2007; 282(3):1539-1543.
123. Paronetto MP, Achsel T, Massiello A et al. The RNA-binding protein Sam68 modulates the alternative splicing of Bcl-x. J Cell Biol 2007; 176(7):929-939.
124. Matter N, Herrlich P, Konig H. Signal-dependent regulation of splicing via phosphorylation of Sam68. Nature 2002; 420(6916):691-695.
125. Navakauskiene R, Treigyte G, Gineitis A et al. Identification of apoptotic tyrosine-phosphorylated proteins after etoposide or retinoic acid treatment. Proteomics 2004; 4(4):1029-1041.
126. Li T, Evdokimov E, Shen RF et al. Sumoylation of heterogeneous nuclear ribonucleoproteins, zinc finger proteins and nuclear pore complex proteins: a proteomic analysis. Proc Natl Acad Sci U S A 2004; 101(23):8551-8556.
127. Vassileva MT, Matunis MJ. SUMO modification of heterogeneous nuclear ribonucleoproteins. Mol Cell Biol 2004; 24(9):3623-3632.
128. Herrmann F, Bossert M, Schwander A et al. Arginine methylation of scaffold attachment factor A by heterogeneous nuclear ribonucleoprotein particle-associated PRMT1. J Biol Chem 2004; 279(47):48774-48779.
129. Liu Q, Dreyfuss G. In vivo and in vitro arginine methylation of RNA-binding proteins. Mol Cell Biol 1995; 15(5):2800-2808.
130. Nichols RC, Wang XW, Tang J et al. The RGG domain in hnRNP A2 affects subcellular localization. Exp Cell Res 2000; 256(2):522-532.
131. Ostareck-Lederer A, Ostareck DH, Rucknagel KP et al. Asymmetric arginine dimethylation of heterogeneous nuclear ribonucleoprotein K by protein-arginine methyltransferase 1 inhibits its interaction with c-Src. J Biol Chem 2006; 281(16):11115-11125.
132. Kim S, Merrill BM, Rajpurohit R et al. Identification of N(G)-methylarginine residues in human heterogeneous RNP protein A1: Phe/Gly-Gly-Gly-Arg-Gly-Gly-Gly/Phe is a preferred recognition motif. Biochemistry 1997; 36(17):5185-5192.
133. Ohkura N, Takahashi M, Yaguchi H et al. Coactivator-associated arginine methyltransferase 1, CARM1, affects pre-mRNA splicing in an isoform-specific manner. J Biol Chem 2005; 280(32):28927-28935.
134. Cheng D, Cote J, Shaaban S et al. The arginine methyltransferase CARM1 regulates the coupling of transcription and mRNA processing. Mol Cell 2007; 25(1):71-83.
135. Sanchez SE, Petrillo E, Beckwith EJ et al. A methyl transferase links the circadian clock to the regulation of alternative splicing. Nature 2010; 468(7320):112-116.
136. Cooper TA, Wan L, Dreyfuss G. RNA and disease. Cell 2009; 136(4):777-793.
137. Miller JW, Urbinati CR, Teng-Umnuay P et al. Recruitment of human muscleblind proteins to (CUG) (n) expansions associated with myotonic dystrophy. EMBO J 2000; 19(17):4439-4448.
138. Kuyumcu-Martinez NM, Wang GS, Cooper TA. Increased steady-state levels of CUGBP1 in myotonic dystrophy 1 are due to PKC-mediated hyperphosphorylation. Mol Cell 2007; 28(1):68-78.
139. Koshelev M, Sarma S, Price RE et al. Heart-specific overexpression of CUGBP1 reproduces functional and molecular abnormalities of myotonic dystrophy type 1. Hum Mol Genet 2010; 19(6):1066-1075.
140. Kanadia RN, Johnstone KA, Mankodi A et al. A muscleblind knockout model for myotonic dystrophy. Science 2003; 302(5652):1978-1980.
141. Ward AJ, Rimer M, Killian JM et al. CUGBP1 overexpression in mouse skeletal muscle reproduces features of myotonic dystrophy type 1. Hum Mol Genet 2010; 19(18):3614-3622.
142. Charlet BN, Savkur RS, Singh G et al. Loss of the muscle-specific chloride channel in type 1 myotonic dystrophy due to misregulated alternative splicing. Mol Cell 2002; 10(1):45-53.

143. Savkur RS, Philips AV, Cooper TA. Aberrant regulation of insulin receptor alternative splicing is associated with insulin resistance in myotonic dystrophy. Nat Genet 2001; 29(1):40-47.
144. Kalsotra A, Xiao X, Ward AJ et al. A postnatal switch of CELF and MBNL proteins reprograms alternative splicing in the developing heart. Proc Natl Acad Sci U S A 2008; 105(51):20333-20338.
145. Ho TH, Charlet BN, Poulos MG et al. Muscleblind proteins regulate alternative splicing. EMBO J 2004; 23(15):3103-3112.
146. Sen S, Talukdar I, Liu Y et al. Muscleblind-like 1 (Mbnl1) promotes insulin receptor exon 11 inclusion via binding to a downstream evolutionarily conserved intronic enhancer. J Biol Chem 2010; 285(33):25426-25437.
147. Paul S, Dansithong W, Kim D et al. Interaction of musleblind, CUG-BP1 and hnRNP H proteins in DM1-associated aberrant IR splicing. EMBO J 2006; 25(18):4271-4283.
148. Osborne RJ, Lin X, Welle S et al. Transcriptional and post-transcriptional impact of toxic RNA in myotonic dystrophy. Hum Mol Genet 2009; 18(8):1471-1481.
149. Iwahashi CK, Yasui DH, An HJ et al. Protein composition of the intranuclear inclusions of FXTAS. Brain 2006; 129(Pt 1):256-271.
150. Jin P, Duan R, Qurashi A et al. Pur alpha binds to rCGG repeats and modulates repeat-mediated neurodegeneration in a Drosophila model of fragile X tremor/ataxia syndrome. Neuron 2007; 55(4):556-564.
151. Sofola OA, Jin P, Qin Y et al. RNA-binding proteins hnRNP A2/B1 and CUGBP1 suppress fragile X CGG premutation repeat-induced neurodegeneration in a Drosophila model of FXTAS. Neuron 2007; 55(4):565-571.
152. Sellier C, Rau F, Liu Y et al. Sam68 sequestration and partial loss of function are associated with splicing alterations in FXTAS patients. EMBO J 2010; 29(7):1248-1261.
153. Morikawa T, Manabe T. Aberrant regulation of alternative pre-mRNA splicing in schizophrenia. Neurochem Int 2010; 57(7):691-704.
154. Aberg K, Saetre P, Jareborg N et al. Human QKI, a potential regulator of mRNA expression of human oligodendrocyte-related genes involved in schizophrenia. Proc Natl Acad Sci U S A 2006; 103(19):7482-7487.
155. Chenard CA, Richard S. New implications for the QUAKING RNA binding protein in human disease. J Neurosci Res 2008; 86(2):233-242.
156. Zhao L, Mandler MD, Yi H et al. Quaking I controls a unique cytoplasmic pathway that regulates alternative splicing of myelin-associated glycoprotein. Proc Natl Acad Sci U S A 2010; 107(44):19061-19066.
157. Chen-Plotkin AS, Lee VM, Trojanowski JQ. TAR DNA-binding protein 43 in neurodegenerative disease. Nat Rev Neurol 2010; 6(4):211-220.
158. Mackenzie IR, Rademakers R, Neumann M. TDP-43 and FUS in amyotrophic lateral sclerosis and frontotemporal dementia. Lancet Neurol 2010; 9(10):995-1007.
159. Lagier-Tourenne C, Cleveland DW. Rethinking ALS: the FUS about TDP-43. Cell 2009; 136(6):1001-1004.
160. Lagier-Tourenne C, Polymenidou M, Cleveland DW. TDP-43 and FUS/TLS: emerging roles in RNA processing and neurodegeneration. Hum Mol Genet 2010; 19(R1):R46-R64.
161. Xiao S, Tjostheim S, Sanelli T et al. An aggregate-inducing peripherin isoform generated through intron retention is upregulated in amyotrophic lateral sclerosis and associated with disease pathology. J Neurosci 2008; 28(8):1833-1840.
162. Rabin SJ, Kim JM, Baughn M et al. Sporadic ALS has compartment-specific aberrant exon splicing and altered cell-matrix adhesion biology. Hum Mol Genet 2010; 19(2):313-328.
163. Lefebvre S, Burlet P, Liu Q et al. Correlation between severity and SMN protein level in spinal muscular atrophy. Nat Genet 1997; 16(3):265-269.
164. Zhang Z, Lotti F, Dittmar K et al. SMN deficiency causes tissue-specific perturbations in the repertoire of snRNAs and widespread defects in splicing. Cell 2008; 133(4):585-600.
165. Cartegni L, Krainer AR. Disruption of an SF2/ASF-dependent exonic splicing enhancer in SMN2 causes spinal muscular atrophy in the absence of SMN1. Nat Genet 2002; 30(4):377-384.
166. Kashima T, Manley JL. A negative element in SMN2 exon 7 inhibits splicing in spinal muscular atrophy. Nat Genet 2003; 34(4):460-463.
167. Young PJ, DiDonato CJ, Hu D et al. SRp30c-dependent stimulation of survival motor neuron (SMN) exon 7 inclusion is facilitated by a direct interaction with hTra2 beta 1. Hum Mol Genet 2002; 11(5):577-587.
168. Bose JK, Wang IF, Hung L et al. TDP-43 overexpression enhances exon 7 inclusion during the survival of motor neuron pre-mRNA splicing. J Biol Chem 2008; 283(43):28852-28859.
169. Chen HH, Chang JG, Lu RM et al. The RNA binding protein hnRNP Q modulates the utilization of exon 7 in the survival motor neuron 2 (SMN2) gene. Mol Cell Biol 2008; 28(22):6929-6938.
170. Singh NN, Seo J, Ottesen EW et al. TIA1 prevents skipping of a critical exon associated with spinal muscular atrophy. Mol Cell Biol 2010.

171. Pedrotti S, Bielli P, Paronetto MP et al. The splicing regulator Sam68 binds to a novel exonic splicing silencer and functions in SMN2 alternative splicing in spinal muscular atrophy. EMBO J 2010; 29(7):1235-1247.
172. Singh NK, Singh NN, Androphy EJ et al. Splicing of a critical exon of human survival motor neuron is regulated by a unique silencer element located in the last intron. Mol Cell Biol 2006; 26(4):1333-1346.
173. Kashima T, Rao N, David CJ et al. hnRNP A1 functions with specificity in repression of SMN2 exon 7 splicing. Hum Mol Genet 2007; 16(24):3149-3159.
174. Kashima T, Rao N, Manley JL. An intronic element contributes to splicing repression in spinal muscular atrophy. Proc Natl Acad Sci U S A 2007; 104(9):3426-3431.
175. Martins de Araujo M, Bonnal S, Hastings ML et al. Differential 3' splice site recognition of SMN1 and SMN2 transcripts by U2AF and U2 snRNP. RNA 2009; 15(4):515-523.
176. Hua Y, Sahashi K, Hung G et al. Antisense correction of SMN2 splicing in the CNS rescues necrosis in a type III SMA mouse model. Genes Dev 2010; 24(15):1634-1644.
177. Hua Y, Vickers TA, Baker BF et al. Enhancement of SMN2 exon 7 inclusion by antisense oligonucleotides targeting the exon. PLoS Biol 2007; 5(4):e73.
178. Hua Y, Vickers TA, Okunola HL et al. Antisense masking of an hnRNP A1/A2 intronic splicing silencer corrects SMN2 splicing in transgenic mice. Am J Hum Genet 2008; 82(4):834-848.
179. Williams JH, Schray RC, Patterson CA et al. Oligonucleotide-mediated survival of motor neuron protein expression in CNS improves phenotype in a mouse model of spinal muscular atrophy. J Neurosci 2009; 29(24):7633-7638.
180. Chu CS, Trapnell BC, Curristin S et al. Genetic basis of variable exon 9 skipping in cystic fibrosis transmembrane conductance regulator mRNA. Nat Genet 1993; 3(2):151-156.
181. Pagani F, Buratti E, Stuani C et al. Splicing factors induce cystic fibrosis transmembrane regulator exon 9 skipping through a nonevolutionary conserved intronic element. J Biol Chem 2000; 275(28):21041-21047.
182. Buratti E, Brindisi A, Pagani F et al. Nuclear factor TDP-43 binds to the polymorphic TG repeats in CFTR intron 8 and causes skipping of exon 9: a functional link with disease penetrance. Am J Hum Genet 2004; 74(6):1322-1325.
183. Buratti E, Dork T, Zuccato E et al. Nuclear factor TDP-43 and SR proteins promote in vitro and in vivo CFTR exon 9 skipping. EMBO J 2001; 20(7):1774-1784.
184. Zuccato E, Buratti E, Stuani C et al. An intronic polypyrimidine-rich element downstream of the donor site modulates cystic fibrosis transmembrane conductance regulator exon 9 alternative splicing. J Biol Chem 2004; 279(17):16980-16988.
185. Dujardin G, Buratti E, Charlet-Berguerand N et al. CELF proteins regulate CFTR pre-mRNA splicing: essential role of the divergent domain of ETR-3. Nucleic Acids Res 2010; 38(20):7273-7285.
186. Venables JP. Unbalanced alternative splicing and its significance in cancer. Bioessays 2006; 28(4):378-386.
187. David CJ, Manley JL. Alternative pre-mRNA splicing regulation in cancer: pathways and programs unhinged. Genes Dev 2010; 24(21):2343-2364.
188. Pettigrew CA, Brown MA. Pre-mRNA splicing aberrations and cancer. Front Biosci 2008; 13:1090-1105.
189. Grosso AR, Martins S, Carmo-Fonseca M. The emerging role of splicing factors in cancer. EMBO Rep 2008; 9(11):1087-1093.
190. Kim MY, Hur J, Jeong S. Emerging roles of RNA and RNA-binding protein network in cancer cells. BMB Rep 2009; 42(3):125-130.
191. Piekielko-Witkowska A, Wiszomirska H, Wojcicka A et al. Disturbed expression of splicing factors in renal cancer affects alternative splicing of apoptosis regulators, oncogenes, and tumor suppressors. PLoS One 2010; 5(10):e13690.
192. He X, Pool M, Darcy KM et al. Knockdown of polypyrimidine tract-binding protein suppresses ovarian tumor cell growth and invasiveness in vitro. Oncogene 2007; 26(34):4961-4968.
193. He Y, Brown MA, Rothnagel JA et al. Roles of heterogeneous nuclear ribonucleoproteins A and B in cell proliferation. J Cell Sci 2005; 118(Pt 14):3173-3183.
194. Venables JP, Klinck R, Koh C et al. Cancer-associated regulation of alternative splicing. Nat Struct Mol Biol 2009; 16(6):670-676.
195. Lapuk A, Marr H, Jakkula L et al. Exon-level microarray analyses identify alternative splicing programs in breast cancer. Mol Cancer Res 2010; 8(7):961-974.
196. LeStourgeon WM, Beyer AL, Christensen ME et al. The packaging proteins of core hnRNP particles and the maintenance of proliferative cell states. Cold Spring Harb Symp Quant Biol 1978; 42 Pt 2:885-898.
197. Christofk HR, Vander Heiden MG, Harris MH et al. The M2 splice isoform of pyruvate kinase is important for cancer metabolism and tumour growth. Nature 2008; 452(7184):230-233.
198. David CJ, Chen M, Assanah M et al. HnRNP proteins controlled by c-Myc deregulate pyruvate kinase mRNA splicing in cancer. Nature 2009.

199. Clower CV, Chatterjee D, Wang Z et al. The alternative splicing repressors hnRNP A1/A2 and PTB influence pyruvate kinase isoform expression and cell metabolism. Proc Natl Acad Sci U S A 2010; 107(5):1894-1899.
200. Patry C, Bouchard L, Labrecque P et al. Small interfering RNA-mediated reduction in heterogeneous nuclear ribonucleoparticule A1/A2 proteins induces apoptosis in human cancer cells but not in normal mortal cell lines. Cancer Res 2003; 63(22):7679-7688.
201. Wang C, Norton JT, Ghosh S et al. Polypyrimidine tract-binding protein (PTB) differentially affects malignancy in a cell line-dependent manner. J Biol Chem 2008; 283(29):20277-20287.
202. Warzecha CC, Sato TK, Nabet B et al. ESRP1 and ESRP2 are epithelial cell-type-specific regulators of FGFR2 splicing. Mol Cell 2009; 33(5):591-601.
203. Warzecha CC, Shen S, Xing Y et al. The epithelial splicing factors ESRP1 and ESRP2 positively and negatively regulate diverse types of alternative splicing events. RNA Biol 2009; 6(5):546-562.
204. Mauger DM, Lin C, Garcia-Blanco MA. hnRNP H and hnRNP F complex with Fox2 to silence fibroblast growth factor receptor 2 exon IIIc. Mol Cell Biol 2008; 28(17):5403-5419.
205. Del Gatto-Konczak F, Olive M, Gesnel MC et al. hnRNP A1 recruited to an exon in vivo can function as an exon splicing silencer. Mol Cell Biol 1999; 19(1):251-260.
206. Carstens RP, Wagner EJ, Garcia-Blanco MA. An intronic splicing silencer causes skipping of the IIIb exon of fibroblast growth factor receptor 2 through involvement of polypyrimidine tract binding protein. Mol Cell Biol 2000; 20(19):7388-7400.
207. Paronetto MP, Cappellari M, Busa R et al. Alternative splicing of the cyclin D1 proto-oncogene is regulated by the RNA-binding protein Sam68. Cancer Res 2010; 70(1):229-239.
208. Valacca C, Bonomi S, Buratti E et al. Sam68 regulates EMT through alternative splicing-activated nonsense-mediated mRNA decay of the SF2/ASF proto-oncogene. J Cell Biol 2010; 191(1):87-99.
209. Ghigna C, Giordano S, Shen H et al. Cell motility is controlled by SF2/ASF through alternative splicing of the Ron protooncogene. Mol Cell 2005; 20(6):881-890.
210. Elliott DJ, Rajan P. The role of the RNA-binding protein Sam68 in mammary tumourigenesis. J Pathol 2010; 222(3):223-226.
211. Matter N, Marx M, Weg-Remers S et al. Heterogeneous ribonucleoprotein A1 is part of an exon-specific splice-silencing complex controlled by oncogenic signaling pathways. J Biol Chem 2000; 275(45):35353-35360.
212. Stark M, Bram EE, Akerman M et al. hnRNP H1/H2-dependent unsplicing of thymidine phosphorylase results in anticancer drug resistance. J Biol Chem 2010.
213. Tauler J, Zudaire E, Liu H et al. hnRNP A2/B1 modulates epithelial-mesenchymal transition in lung cancer cell lines. Cancer Res 2010; 70(18):7137-7147.
214. Lee DH, Chung K, Song JA et al. Proteomic identification of paclitaxel-resistance associated hnRNP A2 and GDI 2 proteins in human ovarian cancer cells. J Proteome Res 2010; 9(11):5668-5676.
215. Shkreta L, Froehlich U, Paquet ER et al. Anticancer drugs affect the alternative splicing of Bcl-x and other human apoptotic genes. Mol Cancer Ther 2008; 7(6):1398-1409.
216. Shkreta L, Michelle L, Toutant J et al. The DNA damage response pathway regulates the alternative splicing of the apoptotic mediator Bcl-x. J Biol Chem 2011.
217. Revil T, Pelletier J, Toutant J et al. Heterogeneous nuclear ribonucleoprotein K represses the production of pro-apoptotic Bcl-xS splice isoform. J Biol Chem 2009; 284(32):21458-21467.
218. Coté J, Dupuis S, Wu JY. Polypyrimidine track-binding protein binding downstream of caspase-2 alternative exon 9 represses its inclusion. J Biol Chem 2001; 276(11):8535-8543.
219. Grammatikakis I, Goo YH, Echeverria GV et al. Identification of MBNL1 and MBNL3 domains required for splicing activation and repression. Nucleic Acids Res 2010.
220. Ule J, Stefani G, Mele A et al. An RNA map predicting Nova-dependent splicing regulation. Nature 2006; 444(7119):580-586.
221. Domsic JK, Wang Y, Mayeda A et al. Human immunodeficiency virus type 1 hnRNP A/B-dependent exonic splicing silencer ESSV antagonizes binding of U2AF65 to viral polypyrimidine tracts. Mol Cell Biol 2003; 23(23):8762-8772.
222. Tange TO, Damgaard CK, Guth S et al. The hnRNP A1 protein regulates HIV-1 tat splicing via a novel intron silencer element. EMBO J 2001; 20(20):5748-5758.
223. Zhou HL, Baraniak AP, Lou H. Role for Fox-1/Fox-2 in mediating the neuronal pathway of calcitonin/calcitonin gene-related peptide alternative RNA processing. Mol Cell Biol 2007; 27(3):830-841.
224. Buratti E, Baralle M, De Conti L et al. hnRNP H binding at the 5' splice site correlates with the pathological effect of two intronic mutations in the NF-1 and TSHbeta genes. Nucleic Acids Res 2004; 32(14):4224-4236.
225. Lin CH, Patton JG. Regulation of alternative 3' splice site selection by constitutive splicing factors. RNA 1995; 1(3):234-245.

226. Zahler AM, Damgaard CK, Kjems J et al. SC35 and heterogeneous nuclear ribonucleoprotein A/B proteins bind to a juxtaposed exonic splicing enhancer/exonic splicing silencer element to regulate HIV-1 tat exon 2 splicing. J Biol Chem 2004; 279(11):10077-10084.

227. Expert-Bezancon A, Sureau A, Durosay P et al. hnRNP A1 and the SR proteins ASF/SF2 and SC35 have antagonistic functions in splicing of beta-tropomyosin exon 6B. J Biol Chem 2004; 279(37):38249-38259.

228. Rooke N, Markovtsov V, Cagavi E et al. Roles for SR proteins and hnRNP A1 in the regulation of c-src exon N1. Mol Cell Biol 2003; 23(6):1874-1884.

229. Crawford JB, Patton JG. Activation of {alpha} -tropomyosin exon 2 is regulated by the SR protein 9G8 and heterogeneous nuclear ribonucleoproteins H and F. Mol Cell Biol 2006.

230. Venables JP, Bourgeois CF, Dalgliesh C et al. Up-regulation of the ubiquitous alternative splicing factor Tra2beta causes inclusion of a germ cell-specific exon. Hum Mol Genet 2005; 14(16):2289-2303.

221. Nasim MT, Chernova TK, Chowdhury HM et al. HnRNP G and Tra2beta: opposite effects on splicing matched by antagonism in RNA binding. Hum Mol Genet 2003; 12(11):1337-1348.

232. Munro TP, Magee RJ, Kidd GJ et al. Mutational analysis of a heterogeneous nuclear ribonucleoprotein A2 response element for RNA trafficking. J Biol Chem 1999; 274(48):34389-34395.

233. Dominguez C, Allain FH. NMR structure of the three quasi RNA recognition motifs (qRRMs) of human hnRNP F and interaction studies with Bcl-x G-tract RNA: a novel mode of RNA recognition. Nucleic Acids Res 2006; 34(13):3634-3645.

234. Matunis MJ, Xing J, Dreyfuss G. The hnRNP F protein: unique primary structure, nucleic acid-binding properties and subcellular localization. Nucleic Acids Res 1994; 22(6):1059-1067.

235. Heinrich B, Zhang Z, Raitskin O et al. Heterogeneous nuclear ribonucleoprotein G regulates splice site selection by binding to CC(A/C)-rich regions in pre-mRNA. J Biol Chem 2009; 284(21):14303-14315.

236. Timchenko LT, Miller JW, Timchenko NA et al. Identification of a (CUG)n triplet repeat RNA-binding protein and its expression in myotonic dystrophy. Nucleic Acids Res 1996; 24(22):4407-4414.

237. Fan XC, Steitz JA. HNS, a nuclear-cytoplasmic shuttling sequence in HuR. Proc Natl Acad Sci U S A 1998; 95(26):15293-15298.

238. Forch P, Puig O, Kedersha N et al. The apoptosis-promoting factor TIA-1 is a regulator of alternative pre mRNA splicing. Mol Cell 2000; 6(5):1089-1098.

239. Le Guiner C, Lejeune F, Galiana D et al. TIA-1 and TIAR activate splicing of alternative exons with weak 5' splice sites followed by a U-rich stretch on their own pre-mRNAs. J Biol Chem 2001; 20:20.

240. Lee JH, Rangarajan ES, Yogesha SD et al. Raver1 interactions with vinculin and RNA suggest a feed-forward pathway in directing mRNA to focal adhesions. Structure 2009; 17(6):833-842.

241. Tsuda K, Kuwasako K, Takahashi M et al. Structural basis for the sequence-specific RNA-recognition mechanism of human CUG-BP1 RRM3. Nucleic Acids Res 2009; 37(15):5151-5166.

242. Charlet BN, Logan P, Singh G et al. Dynamic antagonism between ETR-3 and PTB regulates cell type-specific alternative splicing. Mol Cell 2002; 9(3):649-658.

243. Ponthier JL, Schluepen C, Chen W et al. Fox-2 splicing factor binds to a conserved intron motif to promote inclusion of protein 4.1R alternative exon 16. J Biol Chem 2006; 281(18):12468-12474.

244. Gorlach M, Burd CG, Dreyfuss G. The determinants of RNA-binding specificity of the heterogeneous nuclear ribonucleoprotein C proteins. J Biol Chem 1994; 269(37):23074-23078.

245. Burd CG, Dreyfuss G. RNA binding specificity of hnRNP A1: significance of hnRNP A1 high-affinity binding sites in pre-mRNA splicing. EMBO J 1994; 13(5):1197-1204.

246. Caputi M, Zahler AM. Determination of the RNA binding specificity of the heterogeneous nuclear ribonucleoprotein (hnRNP) H/H'/F/2H9 family. J Biol Chem 2001; 276(47):43850-43859.

247. Amir-Ahmady B, Boutz PL, Markovtsov V et al. Exon repression by polypyrimidine tract binding protein. RNA 2005; 11(5):699-716.

248. Tong A, Nguyen J, Lynch KW. Differential expression of CD45 isoforms is controlled by the combined activity of basal and inducible splicing-regulatory elements in each of the variable exons. J Biol Chem 2005; 280(46):38297-38304.

249. Datar KV, Dreyfuss G, Swanson MS. The human hnRNP M proteins: identification of a methionine/ arginine-rich repeat motif in ribonucleoproteins. Nucleic Acids Res 1993; 21(3):439-446.

250. Kiesler E, Hase ME, Brodin D et al. Hrp59, an hnRNP M protein in Chironomus and Drosophila, binds to exonic splicing enhancers and is required for expression of a subset of mRNAs. J Cell Biol 2005; 168(7):1013-1025.

251. Lerga A, Hallier M, Delva L et al. Identification of an RNA binding specificity for the potential splicing factor TLS. J Biol Chem 2001; 276(9):6807-6816.

252. Rossoll W, Kroning AK, Ohndorf UM et al. Specific interaction of SMN, the spinal muscular atrophy determining gene product, with hnRNP-R and gry-rbp/hnRNP-Q: a role for SMN in RNA processing in motor axons? Hum Mol Genet 2002; 11(1):93-105.

253. Blanc V, Navaratnam N, Henderson JO et al. Identification of GRY-RBP as an apolipoprotein B RNA-binding protein that interacts with both apobec-1 and apobec-1 complementation factor to modulate C to U editing. J Biol Chem 2001; 276(13):10272-10283.
254. Fackelmayer FO, Dahm K, Renz A et al. Nucleic-acid-binding properties of hnRNP-U/SAF-A, a nuclear-matrix protein which binds DNA and RNA in vivo and in vitro. Eur J Biochem 1994; 221(2):749-757.
255. Reimann I, Huth A, Thiele H et al. Suppression of 15-lipoxygenase synthesis by hnRNP E1 is dependent on repetitive nature of LOX mRNA 3'-UTR control element DICE. J Mol Biol 2002; 315(5):965-974.
256. Thisted T, Lyakhov DL, Liebhaber SA. Optimized RNA targets of two closely related triple KH domain proteins, heterogeneous nuclear ribonucleoprotein K and alphaCP-2KL, suggest distinct modes of RNA recognition. J Biol Chem 2001; 276(20):17484-17496.
257. Buckanovich RJ, Darnell RB. The neuronal RNA binding protein Nova-1 recognizes specific RNA targets in vitro and in vivo. Mol Cell Biol 1997; 17(6):3194-3201.
258. Yang YY, Yin GL, Darnell RB. The neuronal RNA-binding protein Nova-2 is implicated as the autoantigen targeted in POMA patients with dementia. Proc Natl Acad Sci U S A 1998; 95(22):13254-13259.
259. Lin Q, Taylor SJ, Shalloway D. Specificity and determinants of Sam68 RNA binding. Implications for the biological function of K homology domains. J Biol Chem 1997; 272(43):27274-27280.
260. Itoh M, Haga I, Li QH et al. Identification of cellular mRNA targets for RNA-binding protein Sam68. Nucleic Acids Res 2002; 30(24):5452-5464.
261. Galarneau A, Richard S. Target RNA motif and target mRNAs of the Quaking STAR protein. Nat Struct Mol Biol 2005; 12(8):691-698.
262. Dejgaard K, Leffers H. Characterisation of the nucleic-acid-binding activity of KH domains. Different properties of different domains. Eur J Biochem 1996; 241(2):425-431.
263. Yeo GW, Coufal NG, Liang TY et al. An RNA code for the FOX2 splicing regulator revealed by mapping RNA-protein interactions in stem cells. Nat Struct Mol Biol 2009; 16(2):130-137.
264. Sauliere J, Sureau A, Expert-Bezancon A et al. The polypyrimidine tract binding protein (PTB) represses splicing of exon 6B from the beta-tropomyosin pre-mRNA by directly interfering with the binding of the U2AF65 subunit. Mol Cell Biol 2006; 26(23):8755-8769.

Chapter 2

The Role of SR and SR-Related Proteins in pre-mRNA Splicing

William F. Mueller and Klemens J. Hertel*

Abstract

Pre-mRNA splicing requires the activities of small nuclear ribonucleoproteins and other essential splicing factors. Among these are members of the SR protein family and SR-related proteins, which are integrally involved in regulating exon recognition, spliceosomal assembly, and spliceosomal re-arrangements to promote intron excision. This chapter will focus on the discovery of SR proteins and SR-related proteins, their structural organization, and their activities in mediating pre-mRNA splicing. The importance of SR-proteins in spliceosomal recruitment will be contrasted to other mechanisms resulting in alternative splicing. An additional aspect of the chapter will be a discussion on how reversible phosphorylation of the RS domain dictates SR protein activity and localization.

Introduction

The splicing of nuclear pre-mRNAs is carried out by the spliceosome, a large dynamic macromolecular complex that recognizes splicing signals and catalyzes the removal of noncoding intronic sequences to assemble protein coding sequences into mature mRNA.[1] A critical step in pre-mRNA splicing is the recognition and pairing of 5' and 3' splice sites. It is possible to evaluate the formation of the mammalian spliceosome in vitro.[2,3] At least four distinct complexes can be resolved by nondenaturing gel electrophoresis in the order of E→A→B→C.[4,5] These complexes differ in composition and order of appearance. Based on these findings, a sequential model of spliceosome assembly was developed. This progression requires the activity of more than 150 distinct protein factors and the U1, U2, U4, U5, and U6 small nuclear RNAs (snRNA) complexed with proteins into small nuclear ribonucleoproteins (snRNPs).[6] The first intermediate of this assembly reaction is E complex, which is characterized by ATP independent, stable interactions of U1 snRNP with the 5' splice site[7-9] and U2 auxiliary factor (U2AF) with the polypyrimidine tract.[7,10,11] ATP hydrolysis then leads to the formation of A complex, which is characterized by the stable association of U2 snRNP with the branch point/3' splice site sequence and functional commitment to splice site pairing.[12,13] B complex and the catalytically active C complex form after the incorporation and re-arrangement of the U4•U6/U5 tri-snRNP and facilitate the excision of the intron and ligation of exons.

History and Discovery

In addition to snRNPs, pre-mRNA splicing requires other protein factors to efficiently remove introns and ligate exons. One class of regulatory proteins indispensible in this process is the SR protein family. Fractionation and purification experiments of proteins associated with spliceosomes

*Department of Microbiology and Molecular Genetics, University of California, Irvine, California, USA.
Corresponding Author: Klemens J. Hertel—Email: khertel@uci.edu

RNA Binding Proteins, edited by Zdravko J. Lorković.
©2012 Landes Bioscience.

in addition to genetic screens in *Drosophila* led to the discovery of SR proteins. Fractions from mammalian cell extracts were used in various inactivation and complementation assays to demonstrate that certain proteins, as well as snRNPs, were necessary for splicing.[14] These studies laid the foundation for the classical complementation studies that continue to be used to characterize splicing factors and, eventually, to identify SRSF1 and SRSF2, the most widely studied of the human SR proteins.[15,16] Sequence analysis of these proteins revealed the presence of extended arginine and serine dipeptides, termed the arginine/serine (RS) domain, in addition to at least one RNA binding domain (RBD) of the RNA Recognition Motif (RRM)-type.[16-18] The SR protein involvement in splicing became more obvious when additional studies in *Drosophila* identified SWAP (*suppressor-of-white-apricot*), Tra (*transformer*),[19] and Tra-2 (*transformer-2*)[20] as promoting specific splicing patterns in mature mRNAs. The identification of these factors, all of which contained RS domains, suggested that RS domain-containing proteins are integrally involved in splice site choice.

The original family of SR proteins was classified following the identification of other RS domain-containing proteins based on: The presence of a phosphoepitope that is recognized by the monoclonal antibody mAb104,[21] their conservation across vertebrates and invertebrates, their ability to precipitate in 20 mM Magnesium Chloride,[21] and their activity in splicing complementation.[22] Later on it was discovered that SR proteins can also function outside of their canonical splicing roles in other aspects of gene expression, such as transcription and mRNA export.[23-25] Due to the increase in activities and the confusing nature of SR protein classification, the definition of an SR protein has recently been altered to only be based on specific sequence properties: "One or two N-terminal RBDs (PF00076), followed by a downstream RS domain of at least 50 amino acids with >40% RS content, characterized by consecutive RS or SR repeats."[26] In humans, the SR protein family is encoded by twelve genes, designated Serine/Arginine-rich Splicing Factor (SRSF) 1-12 (Table 1). All twelve members of the SR protein family have a common structural organization and fulfill the domain requirement of the definition (Fig. 1).[26,27]

The classification of SR proteins excludes many proteins that contain RS domains but do not meet the other SR protein classification criteria. These proteins are referred to as SR-related proteins. Bioinformatic surveys have found multiple proteins that contain RS domains,[25,28,29] but may have different or no RNA binding domains and may lack the ability to complement splicing reactions (Table 2, Fig. 2). These SR-related proteins are less highly conserved but still may interact with the spliceosome and alter splicing activity.[29-31] SR-related proteins function in multiple RNA processing pathways, expanding the SR protein family and increasing the complexity with which splicing can be regulated.[32] By differential binding to RNA elements and differential interactions with other splicing factors through RS domain interactions, SR family proteins influence the way introns and exons are recognized and spliced.

While introns are common to eukaryotes, the complexity of alternative splicing varies among species. Accordingly, SR proteins exist in all metazoan species[22] as well as in some lower eukaryotes, such as *Schizzosaccharomyces pombe*.[33,34] However, classical SR proteins are not present in all eukaryotes and are apparently missing from *Saccharomyces cerevisiae*, which also lacks alternative splicing. Instead, three SR-like proteins have been identified in *S. cerevisiae*, one of which, Npl3, has been shown to modulate the efficiency of splicing.[35] Of the SR-related proteins found in higher eukaryotes, some (mostly snRNP associated proteins) have orthologs in yeast that lack RS domains, but many lack yeast orthologs all together.[29] In general, the species-specific presence of SR family proteins correlates with the presence of RS domains within other components of the general splicing machinery allowing interaction between them. The observation that the density of RS repeats correlates with the conservation of the branch-point signal, a critical sequence element of the 3' splice site, argues for an evolutionary origin of SR proteins.[36] As such, SR family proteins appear to be ancestral to eukaryotes, but were independently lost in some lineages. Phylogenetic tree analyses further suggest that successive gene duplications played an important role in SR protein evolution.[37] These duplication events are coupled with high rates of nonsynonymous substitutions that promoted positive selection favoring the gain of new functions, supporting the hypothesis that the expansion of RS repeats during evolution had a fundamental role in the relaxation of the splicing signals and in the evolution of regulated alternative splicing.

Table 1. List of SR proteins, their current and previous nomenclature, RNA consensus binding sequences, and functions

SR Protein	Previous Name	Consensus Binding Sequence	Function	Reference
SRSF1	ASF/SF2, SRp30a	RGAAGAAC, AGGACRRAGC, SRSASGA, UGRWG	Constitutive and alternative splicing	51, 56, 204
SRSF2	SC35, SRp30b	AGSAGAGUA, GUUCGAGUA, UGUUCSAGWU, GWUWCCUGCUA, GGGUAUGCUG, GAGCAGUAGKS, AGGAGAU, GRYYCSYR	Constitutive and alternative splicing	52, 56, 81, 82
SRSF3	SRp20	GGUCCUCUUC WCWWC CUCKUCY	Constitutive and alternative splicing	81, 82, 133
SRSF4	SRp75	GAAGGA	Constitutive and alternative splicing	149
SRSF5	SRp40	GAGCAGUCGGCUC, ACDGS	Constitutive and alternative splicing	51, 176
SRSF6	SRp55	USCGKM, UCAACCAGGGGAC	Constitutive and alternative splicing	51, 92
SRSF7	9G8	UCAACA ACGAGAGAY GGACGACGAG	Constitutive and alternative splicing	81, 82, 159
SRSF8	Human SRp46	Not determined	Constitutive and alternative splicing	139
SRSF9	SRp30c	GACGAC, AAAGAGCUCGG, CUGGAUU	Constitutive and alternative splicing	65, 140
SRSF10	SRp38, SRrp40, hTra2Beta	(GAA)n	Constitutive and alternative splicing	93
SRSF11	SRp54, p54	C rich regions	General splicing repressor	104
SRSF12	SRrp35, SRp86	Not determined	Negative of alternative splicing	148

N: any amino acid; Y: pyrimidine; R: purine; D: A,G,U; K: U,G; M: A,C; S: G,C; W: A,U. SR and SR-related tables are reviewed in references 25, 27 and 29.

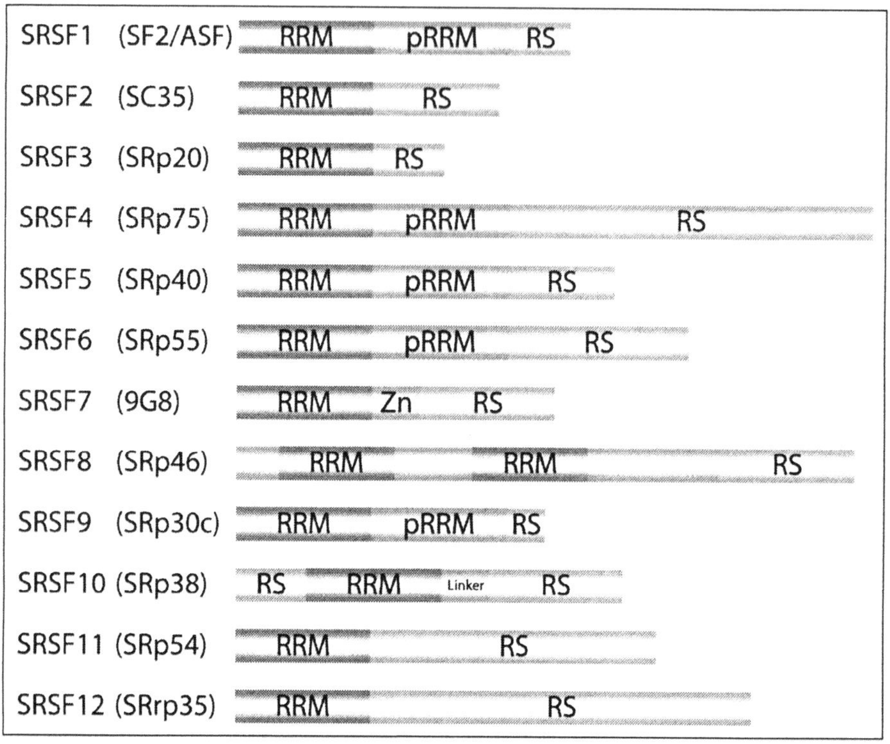

Figure 1. Domain organization of SR-proteins. SR proteins identified by their most recent nomenclature (left column), followed by their previous nomenclature (middle column), followed by a cartoon representation of their major domains: RNA Recognition Motifs (RRM), Pseudo RRM domains (pRRM), RS domains (RS), Zinc finger domains (Zn) and linker regions.

Table 2. List of selected SR-related proteins, their current nomenclature and functions

SR Related Protein	Function	Reference
U170K	Splicing factor	68, 70
U2AF65	Splicing factor	11
U2AF35	Splicing factor	153
U5100K	Splicing factor	43
DDX46	Spliceosomal re-arrangement	151
ClkSty1-3	SR protein kinase	150
Urp	Splicing factor	158
SRm160	Splicing activator	152
SRm300	Splicing activator	30

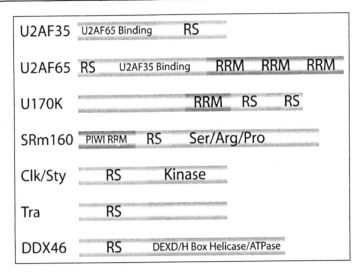

Figure 2. Domain organization of representative SR-related proteins. The cartoon highlights the major domains found in each protein: RNA Recognition Motifs (RRM), RS domains (RS), protein interaction domains, kinase domains, helicase/ATPase domains, and linker regions (unlabelled). This figure highlights the diversity of the SR-related proteins and their ubiquitous nature throughout splicing.

Structural Elements and Binding Specificities of SR and SR-Related Proteins

All SR proteins share two main structural features: The RS domain and at least one RRM (Fig. 1). It is believed that RS domains facilitate protein-protein and protein-RNA interactions and by doing so mediate spliceosomal recruitment, assembly, and function (discussed in the SR family activity section).[10,38-41] The RS domains of SR family proteins participate in protein interactions with a number of RS domain containing splicing factors.[38,42] These factors include other SR proteins, SR-related proteins, and components of the general splicing machinery.[29,38,42-45] Furthermore, the RS domain interacts with RNA[46] and it can function as a nuclear localization signal by mediating the interaction with nuclear import receptor transportin-SR.[47-49] These activities can be modulated through post-translational modifications, mainly via phosphorylation (discussed in the SR family activity section).[29,31]

The presence of RRM domains is the other structural feature defining SR proteins. For the majority of SR proteins with two RNA binding domains, the second domain is a poor match to the RRM consensus and is referred to as an RRM homolog called a pseudo RRM (pRRM). The only exception is SRSF7, which contains an RRM and a zinc-knuckle domain that is thought to contact the RNA.[50] In the cases where it has been determined, SR proteins have specific, yet degenerate RNA binding specificities.[51,52] Originally, it was shown that SRSF1 and 2 bind to 5', and 3' splice sites to facilitate U1 snRNP, U2AF, and U2 snRNP interactions with the splice sites and with each other in E complex.[18,53-55] However, this binding was also shown to be low affinity binding[56] suggesting that there were better targets for SR proteins outside of splice sites. Soon after this, purine-rich consensus sequences were found that displayed higher affinity binding for SRSF1 and 2.[56] These purine-rich sequences were discovered through mutational analysis as exonic splicing enhancers (ESEs), exonic sequences that activate specific exon inclusion.[57-59] It was soon found that SR proteins promiscuously interact with multiple different ESEs,[59-65] demonstrating that they have degenerate binding specificity (see Table 1 and Splicing activity of SR proteins section). More recently, chromatin immunoprecipitation experiments were used to show that SR proteins are recruited cotranscriptionally and that the site of recruitment to the nascent RNA is mainly dictated by their RRM interactions with the pre-mRNA.[66]

SR-related proteins have only an RS domain to define them (Fig. 2). Many SR-related proteins lack RRMs all together and instead have domains that mediate interactions with other cellular machineries or contain domains that modify other RNA processing machineries (such as kinases with RS domains).[29,31] In the case of the SR-related protein U1-70K, the small RS domain is conserved from *Drosophila* to humans, but not in yeast.[67-71] The RRM of U1-70K binds to U1 snRNA through a conserved 8 amino acid sequence that was found in multiple RNA binding proteins.[72] U1 snRNP binds to 5′ splice sites through RNA/RNA and protein/RNA interactions aided by SR proteins through U1-70K's RS domain. Additionally, U1-70K's RS domain assists in bridging splice sites during early spliceosomal steps.[38,42] Despite recent structural analyses revealing extensive interaction throughout U1 snRNP, the RS domain binding abilities have yet to be characterized structurally.[73,74]

For 3′ splice site recognition, the 65 and 35 kD subunits of U2AF mediate polypyrimidine tract binding and subsequent interaction with U2 snRNP.[38,75-77] Through their interaction with the pre-mRNA, both of these essential SR-related proteins allow for snRNP and other SR protein recruitment. In an interaction reminiscent of U2AF 35 and 65 scenario, the *Drosophila* Tra protein, which has no RRM, must interact with Tra 2, which has an RRM, to bind to the *doublesex* pre-mRNA.[78] Other RS-related proteins contain DEXD/H-box ATPase/helicase domains, CTD interacting domains, kinase domains, and many others, all important in modulating splicing activity.[29]

Few SR-related or SR proteins have been characterized structurally, mainly due to solubility issues, probably involving the large exposed hydrophobic regions. Unfortunately, no structural information detailing the RS domain is available to date. This may be explained by the poor solubility of these proteins in their free state and the unknown phosphorylation state of the serines within the RS domain. In addition, the degenerate RNA-binding sequences recognized by SR proteins may have prevented their study in the bound form. Consequently, only isolated RRMs of SR proteins have been analyzed structurally by nuclear magnetic resonance spectroscopy. To tackle the solubility issues, the RRMs of SRSF3 and SRSF7 were fused to the immunoglobulin G-binding domain 1 of Streptococcal protein G (GB1) solubility tag[79] or overexpressed RRMs were suspended in a solution containing charged amino acids.[80] Using these manipulations it was possible to obtain solution structures of the free SRSF7 and SRSF3 RRMs and of the SRSF3 RRM in complex with the RNA sequence 5′-CAUC-3′. When analyzing the unbound RRMs of SRSF3 and SRSF7, one is struck by an unusually large exposed hydrophobic surface, which could explain why the solubility of SR proteins is so low. The SRSF3 RRM complex with RNA shows that although all four nucleotides present are contacted by the RRM only the 5′ cytosine is recognized in a specific manner. These structural insights provided a potential explanation for the seemingly low RNA binding specificity exhibited by SRSF3.[81,82]

RS domain activities are regulated through phosphorylation, and thus require kinase interaction (see the SR protein regulation section). In some cases this could be mediated through RS domains found in certain SR protein kinases (Clk/Sty proteins 1-3), but in the case of SRPK1, interactions are established through a unique docking groove to SRSF1 directing phosphorylation to specific areas of its RS domain. This docking groove seems to cancel out the charges on the RS domain with opposing charges allowing deep and tight binding and, eventually, allowing processive phosphorylation.[83-86] These observations suggest a structurally related regulatory mechanism for phosphorylation of RS domains, allowing specific kinases to recognize specific RS domains based on specific docking motives. However, due to the problems inherent to crystallizing RS domain containing proteins, it seems that more data regarding how RS domains interact will be long in coming.

Splicing Activity of SR Proteins

Splicing Activation

For classical cases of alternative splicing, it was shown that cis-acting RNA sequence elements increased exon inclusion by serving as binding sites for the assembly of multi-component splicing enhancer complexes. For example, a specific exonic sequence recruited Tra, Tra 2, and RBP1 (the *Drosophila* homolog of SRSF3) to form a heterotrimeric complex regulating sex specific alternative splicing of the *Drosophila doublesex* gene.[60] Additional studies suggested that the RRMs directed SR proteins to sequences on the pre-mRNA through which they could interact with other proteins and influence splicing.[59,62,87] These RNA sequence elements were usually located within the regulated exon leading to their definition as ESEs.[1,88] In general, ESEs are recognized by at least one member of the SR protein family and recruit the splicing machinery to the adjacent intron.[1,27,88] This recruitment has been shown to activate weaker splice sites or to alter splice site choice.[15,16] Surprisingly, SR protein binding sites are not only limited to alternatively spliced exons, but they have also been verified for exons of constitutively spliced pre-mRNAs.[89,90] It is therefore likely that SR proteins bind to sequences found in most, if not all, exons and are thus active in the greater majority of all exon recognition.

It was soon discovered that the same SR proteins could display slightly different binding sequences in different contexts, thus new methods were established for finding ESEs. Using a modified binding SELEX technique,[91] degenerate consensus RNA sequences were identified that interact with SRFS1 and 2.[56] Subsequent enrichment approaches were based on the ability of an SR protein to activate intron removal.[51] These functional SELEX techniques were used to generate binding consensus sequences for all SR proteins and for some SR-related proteins.[81,82,89,92,93] In combination with the deciphering of the human genome, this information allowed global computational analyses of pre-mRNA sequences to determine the density of ESEs within the genome.[94,95] The results from these bioinformatic approaches showed that potential SR protein binding sites are found in far more places than previously thought, supporting the notion that SR protein activity is paramount for splice site selection throughout the human genome.

Using recombinant SR proteins to complement splicing deficient cell extracts it was demonstrated that SR proteins functionally overlapped in many cases.[22,50,96,97], suggesting that some SR proteins may be functionally redundant. However, despite many similarities, SRSF1 and 2 cannot replace each other and they bind to different ESEs.[50,98] RS domain swapping experiments further displayed different activation potentials or even splicing repression (see also the Splicing Repression section below).[99] The strongest argument for nonredundant functions in vivo originate from the early embryonic lethal phenotypes of SR protein knock out mice, clearly demonstrating the developmental significance of these splicing factors in different pathways.[25]

SR proteins facilitate the binding of regulatory complexes to the pre-mRNA and promote splicing through mediating the recruitment of prespliceosomal or E complex components such as U1 snRNP, and the proteins SF1 and the U2AF heterodimer (Fig. 3A). Multiple SR proteins have been shown to increase the kinetics of E complex formation.[42,100] It was even found that the need for U1 snRNP binding to the 5' splice site could be bypassed in the presence of excessive amounts of recombinant SR proteins.[101,102] In addition to their enhancement of splice site recognition, it was demonstrated that SR proteins bridge across introns or exons to facilitate splice site pairing[18,38] through interacting with components from both U1 and U2 snRNPs.[103,104] These splice site pairing interactions can be recapitulated in trans-splicing reactions, in which SR proteins also play an essential role.[55,105-109]

The action of SR-mediated splice site recognition does not go unchecked. In addition to splicing enhancers, sequences have been identified that decrease exon inclusion efficiency (referred to as splicing silencers). In general, hnRNPs or other similar splicing inhibitory proteins recognize these RNA elements.[110-112] Competition between the activities of SR proteins and hnRNP A1 have been implicated in the regulation of HIV TAT exon 3[113] as well as IgG exons 1 and 2[77] pointing

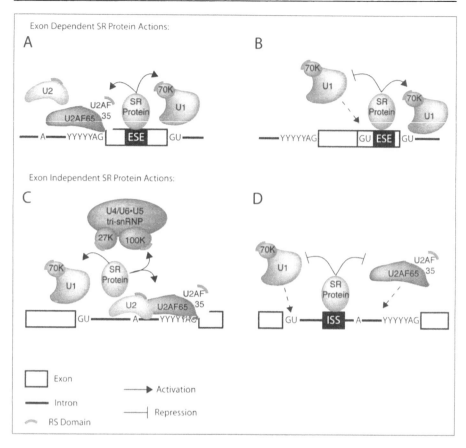

Figure 3. Pre-mRNA splicing activities of SR proteins. A) SR proteins assist in the recruitment of spliceosomal complexes to the pre-mRNA through RRM interaction with the RNA and RS domain interaction with other proteins. SR-related proteins can display equivalent recruitment functions (not shown). B) Directional activity of SR proteins. SR proteins can simultaneously exhibit activating and inhibitory functions based on their orientation relative to competing splice sites. Because these activities require interactions of SR proteins with the regulated exon (A, B), they are referred to as exon-dependent actions of SR proteins.[51-128,134-149] C) SR mediated recruitment of the tri-snRNP and stabilization of previously recruited snRNPs. These interactions are essential for spliceosomal re-arrangements and proper function. SR-related proteins such as SRm160/300, DDX46, and others function similarly (not shown). D) SR-mediated splicing repression can occur when SR proteins bind with high affinity to the intron. The precise mechanism of this inhibition is unknown. Because these SR protein activities do not require interactions with the regulated exon (C, D),[129-132] they are referred to as exon-independent actions of SR proteins.[1,15,16,27]

to a yin/yang model of splicing regulation. While SR proteins generally promote the formation of functional spliceosomes, hnRNPs or other proteins generally act against exon recognition. Interestingly, the role of SR proteins as activators or as inhibition blockers seems to change depending on the type of splicing occurring, alternative or constitutive. In vitro tests of mutant SR proteins have shown that the RS domain is dispensable for certain alternative splice site selection events, but necessary for constitutive splicing.[114-116] This data supports the idea that SR proteins recruit the spliceosome to activate alternative splice sites via their RS domains, and that they also inhibit repressive splicing mechanisms through the substrate specific binding of their RRMs.

These two modes of activity are not mutually exclusive as evidence points to SR proteins having multiple context dependent roles.[117]

An alternative mode of spliceosomal recruitment was suggested by experiments demonstrating that the RS domains of U2AF65 and SRSF1 contact the branch point and 5′ splice site in the mRNA. As these contacts are established within the functional spliceosome, it is likely that SR family proteins contact the pre-mRNA throughout the splicing process.[40,41] Modulation of RNA binding by the RS domain seems to make SR protein binding less specific and more promiscuous.[51] Irrespective of the RS domain activation mode, SR proteins facilitate the recruitment of spliceosomal components to the regulated splice site.[1,118] Thus, SR proteins bound to ESEs function as general activators of exon definition (Fig. 3A).[119] Kinetic analyses demonstrated that the relative activity of ESE-bound SR proteins determine the magnitude of splicing promotion. This activity depended on the number of SR proteins assembled on ESEs and the distance between ESEs and the intron. It was also shown that splicing activation was proportional to the number of serine-arginine repeats contained within the RS domain of the bound SR protein. Thus, the quantity of serine-arginine repeats appears to dictate the activation potential of SR proteins.[120]

In cases of alternative splicing, the activation by SR proteins was found to promote the pairing of the most proximal splice sites across the intron (Fig. 3B).[15,121-123] Interestingly, SR proteins do not do this by simply enhancing recognition of the proximal splice site alone, they also inhibit the use of the distal splice site through yet unknown mechanisms.[117,124] This directional activation of splice sites seems to depend on splice site strength and is probably modulated by the presence of other enhancing or silencing signals (see also Splicing Repression section below).[125-127] Even in the presence of SR proteins, neighboring splice sites play a significant role during splice site selection, as demonstrated in cases where a nearby 5′ splice site increases the usage of another 5′ splice site.[117,128]

Exon Independent Activation

Characterization of SRSF1's activity showed that 5′ splice site cleavage, branch point choice, and lariat formation were all affected by its presence.[15,16] These observations pointed to SR protein activity in more than just the initial splice site recognition and pairing steps of splicing. Similar results were obtained with other SR proteins.[129,130] Mechanistic hints as to why SR protein action is important for later spliceosomal assembly steps came soon after it was found that the U4/U6•U5 tri-snRNP required SR proteins to be incorporated into the spliceosome (Fig. 3C).[131] Because tri-snRNP incorporation succeeds initial splicing steps, no exonic sequences were found to be specifically required. In fact, SR proteins have essential activities that do not require interactions with exon sequences to initiate the first step of splicing.[132] The role of the exon-independent function may be to promote the pairing of 5′ and 3′ splice sites across the intron or exon,[109] facilitate cross talk between RNA processing events[133] and/or to facilitate the incorporation of the U4/U6•U5 tri-snRNP into the spliceosome.[131] While the RRM is essential for its exon-independent activity,[132] it is likely that SR proteins interact with the partially assembled spliceosome or the tri-snRNP through RS domain contacts.

Splicing Repression

One striking feature of SR proteins is their prevalent location within the pre-mRNA. In nearly all cases SR proteins have been found to interact with exonic sequences of the pre-mRNA. This is a surprising finding considering the fact that their relatively promiscuous binding specificity predicts that introns are littered with potential SR protein binding sites. The fact that SR proteins are mainly observed to bind within exonic sequences suggests that additional requirements need to be met for functional SR protein binding to the pre-mRNA. However, some SR proteins do bind within the intron where they generally function as negative regulators of splicing (Fig. 3D). The best-characterized example of this is observed during adenovirus infection.[134] In this case, splicing is repressed by the binding of the SR protein SRSF1 to an intronic repressor element located upstream of the 3′ splice site branchpoint sequence in the adenovirus pre-mRNA. When bound to the repressor element, SRSF1 prevents the recruitment of U2 snRNP to the branchpoint sequence, thereby inactivating the 3' splice site. Other studies provided further support for the idea that SR proteins placed within introns generally interfere with the productive assembly of spliceosomes.[117,135-137] SRSF2 interacts with an intronic sequence of the _Beta-tropomyosin_ gene antagonizing the action of SRSF1 in the recognition of exon 6A.[138] These observations suggest that expression levels of SR proteins may be very important in regulating splicing decisions, as the balance between competing SR protein binding sites, exonic and intronic, will affect the splicing outcome.[139,140]

Some SR proteins seem to generally exhibit inhibitory actions or consistently oppose the "canonical" SR protein functions. SRSF11 competes with human Tra2β to promote exon skipping and distal splice site activation.[99,141] In addition, SRSF5,10, and 12 have also been implicated in negative regulation of splicing, although SRSF10 acts only after dephosphorylation in response to mitosis or cellular stress (see SR protein regulation below).[142-145] SRSF12 was found to interact with all the SR proteins in both enhancing and suppressing contexts. Curiously, while SRSF12 modulates SR protein function, its activity may also be modified by interactions with SR proteins and hnRNPs.[146-148] These observations demonstrated that SR proteins, through context dependent binding of RNA and context dependent protein interactions, act to facilitate spliceosomal complex formation, as barriers to prevent exon skipping, or as inhibitors of splice site usage.[149]

SR-Related Protein Activity

Due to the multitude of RS domain containing proteins, actions of SR-related proteins influence many cellular processes (Fig. 4). They are involved in bridging splicing reactions, regulation of other RS domain containing proteins, and are components in the main splicing machinery itself.[25,29,150,151] Even though they are not studied as extensively as SR proteins, it is appreciated that their functions in splicing are important and, in many cases, necessary.[152]

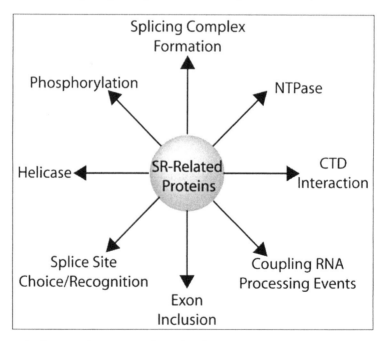

Figure 4. The functional repertoire of SR-related proteins in pre-mRNA splicing. SR-related proteins have been shown to participate in spliceosomal complex formation, phosphorylation of RS domains, splice site recognition, exon inclusion and the coupling of RNA processing events (transcription, polyadenylation, capping, nuclear export). In addition, some SR-related proteins have essential helicase or ATPase functions.[150-168]

Like SR proteins, SR-related proteins act early in splicing to aid exon recognition and complex formation. The best studied of these early acting splicing proteins are components of the early spliceosomal complexes themselves: U1-70k and both subunits of U2AF. Multiple SR proteins interact with these SR-related proteins to promote their recruitment to splice sites and interaction across introns and exons.[11,38,42,75,153] U1 snRNP has been shown to stabilize pre-mRNAs and to influence splice site choice.[117,154] U2AF competes for pre-mRNA binding with other polypyrimidine tract binding proteins that interfere with spliceosome assembly.[155,156] Interestingly, multiple genes exist that have a high similarity to U2AF. Presumably, these U2AF related proteins function in subsets of pre-mRNA splicing, providing unique ways to identify and interact with the 3' splice site.[157,158]

The SR-related proteins SWAP, Tra, and Tra 2, were among the first proteins implicated in alternative splicing regulation and their human homologues have been shown to have similar activities.[159-162] SRm160/300 is a large matrix protein that interacts with multiple RNA interaction proteins with multiple RS domains and contains a PWI RNA recognition domain.[30,163,164] It is critical for bridging splice sites on certain pre-mRNAs and has been found to complex with transcription related proteins and chromatin support proteins.[165] These observations suggest a mechanistic link between splicing and transcription.[166]

Some SR-related proteins have also been shown to regulate SR protein activity by directing their phosphorylation via kinase domains. One of the best characterized of these are the Clk/Sty protein kinases. Clk/Sty 1 interacts with SR proteins and phosphorylates their RS domains, thus altering their activity and localization (see the SR protein regulation section).[166,167] Similarly, the hPrp4 kinase interacts with components of the tri-snRNP and phosphorylation of U4/U6 associated PRP31 and U5 associated PRP6 is necessary for successful transition into spliceosomal B complex.[108] The U4/U6•U5 tri-snRNP contains many more RS domain containing proteins, all of which are potential targets for similar regulation mechanisms.[43,44,168]

SR Protein Regulation and Localization

SR proteins are expressed throughout development in most tissues and organs. To control the production of the appropriate transcriptome, SR proteins are regulated at the translational and post-translational levels, either through self-regulation (Sxl and SWAP), or through interactions with other SR family proteins. In all of these cases, the control over SR protein expression is necessary for maintaining efficient splicing in all cell types.

Like other proteins involved in pre-mRNA splicing, SR proteins are enriched in nuclear compartments, termed speckles. Speckles consist of two distinct structures: Interchromatin granule clusters (IGCs), storage/re-assembly sites for pre-mRNA splicing factors that are 20-25 nm in diameter, and perichromatin fibrils (PFs), the site of actively transcribing genes and cotranscriptional splicing that are approximately 5 nm in diameter.[169,170] The SR proteins are one prominent component of nuclear speckles[31,171] and biochemical analyses have indicated that RS domains are responsible for targeting the SR family proteins to speckles and that this localization can be slightly different for each SR protein.[47,172] Because the nuclear organization of SR proteins is dynamic, SR proteins are recruited from IGC storage clusters to the site of cotranscriptional splicing (PFs).[173,174] Interestingly, both the RNA binding domains and RS domains are required for the recruitment of SR proteins from IGCs to PFs, as is phosphorylation of the RS domain.[175,176]

SR proteins act at several steps during the splicing reaction[16,17,22,98,131,177] and require phosphorylation and dephosphorylation for spliceosomal assembly.[178,179] It was found that hyper- or hypo- phosphorylation inhibited splicing.[166,167] This is evident in the case of SRSF1 where phosphorylation is required for interactions with U1-70k, yet dephosphorylation is required for splicing to continue to catalysis.[180-182] A number of SR protein kinases have been shown to specifically phosphorylate serine residues within the RS domain of SR proteins. This seems to occur in a temporal and region specific manner, i.e., certain kinases prefer to act on certain areas of the RS domain with different specificities at different times. These include SR

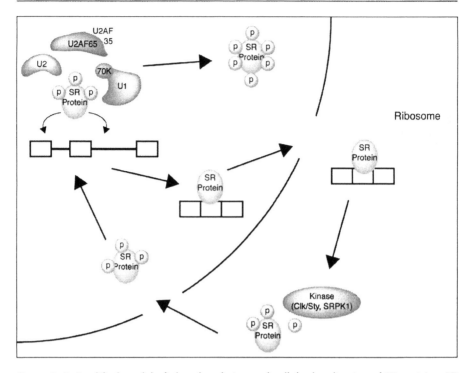

Figure 5. A simplified model of phosphorylation and cellular localization of SR proteins. SR proteins require phosphorylation to mediate spliceosomal complex formation. However, dephosphorylation is required for splicing catalysis to occur, as well as nuclear export. Re-entry into the nucleus requires rephosphorylation where the SR protein can act again in the splicing pathway. As most SR proteins shuttle between the nucleus and cytoplasm, this cycle may provide a simple mode for functional regulation. Dephosphorylation of SR proteins leads to sequestration to the cytoplasm, altering SR protein levels in the nucleus and changing splicing patters. By contrast, hyperphosphorylation of SR proteins represses splicing.[169-198]

protein kinase 1 (SRPK1),[183] Clk/Sty kinase (see the SR related protein activity section),[166] cdc2p34[184] and topoisomerase.[185] SRPK1 has been crystallized in complex with the docking domain of SRSF1 and seems to preferentially phosphorylate the N-terminal segment of the RS domain.[84] This region-specific phosphorylation is contrasted against the Clk/Sty kinase that acts on the whole RS domain.[83] These two examples allow a simplified view of how SRSF1 may be regulated (Fig. 5).

It has been shown that SR proteins may be regulated by their localization with in the cell. The phosphorylation state of the RS domain triggers the shuttling of SR proteins between the cytoplasm and the nucleus (Fig. 5).[47,186] From their location in nuclear speckles, phosphorylated SR proteins perform their functions in splicing after which a subset of SR proteins are active in mRNA export as adaptor molecules.[66,187-190] This adaptor function requires dephosphorylation, followed by export along with the mature mRNA.[186,189,191] Rephosphorylation is required for these shuttling SR proteins to return to the nucleus to carry out their nuclear splicing functions (Fig. 5). Several kinases found in the cytoplasm are responsible for this process, supporting an SR protein localization cycle that is intimately linked to mRNA production and the cell cycle.[192-194] Unique pathways of this cycling exist in response to heat shock,[144,195] cellular growth,[196] and signaling[197] to change SR protein concentrations within the nucleus, thus modulating mRNA splice patterns.[198]

Conclusion

The SR protein family engages in ubiquitous and promiscuous interactions throughout the splicing pathway. Understanding of the action and regulation of these proteins seems daunting in face of all their cellular activities, which extends beyond those mentioned in this chapter. Global analyses of SR proteins have fueled the ability to predict potential binding sites of SR proteins.[95,199-202] While the resulting computational tools have helped to suggest location from which SR proteins function, it has also lead to an embarrassment of riches, with SR binding sequences found frequently throughout the genome. When these results are compared with the most recent attempts to understand splicing through an analysis of cis-acting elements, the 'splicing code', it appears that the presence of SR protein binding sites are not as important to exon inclusion as many other RNA elements investigated.[203] Using CLIP-seq techniques it was further demonstrated that some programs to predict SR protein binding sites may be over ambitious.[204] It is clear that SR proteins are necessary for accurate splicing to occur, presumably within a subset of unique pre-mRNAs. The challenge to further understand the actions of this diverse family of proteins will rest on the continued analysis of actual binding sites using genome-wide techniques. This sort of global analysis may help to define different situations where SR proteins are most active and important. While the current understanding of this family of proteins is taking a shape that allows sufficient understanding of many splicing and splicing-related disease phenotypes, there is still a great amount of work to be done to create a more complete picture of their activities and regulation during pre-mRNA splicing.

Acknowledgments

The authors wish to acknowledge support from the NIH (RO1 GM62287 and R21 CA149548 to K.J.H.).

References

1. Black DL. Mechanisms of alternative pre-messenger RNA splicing. Annu Rev Biochem 2003; 72:291-336.
2. Will CL, Lührmann R. Protein functions in pre-mRNA splicing. Curr Opin Cell Biol 1997; 9(3):320-328.
3. Reed R, Palandjian L. Spliceosome assembly. In: Krainer AR, ed. Eukaryotic mRNA Processing. Oxford: IRL Press; 1997:103-129.
4. Konarska MM, Sharp PA. Electrophoretic separation of complexes involved in the splicing of precursors to mRNAs. Cell 1986; 46(6):845-855.
5. Das R, Zhou Z, Reed R. Functional association of U2 snRNP with the ATP-independent spliceosomal complex E. Mol Cell 2000; 5(5):779-787.
6. Jurica MS, Moore MJ. Pre-mRNA splicing: awash in a sea of proteins. Mol Cell 2003; 12(1):5-14.
7. Bennett M, Michaud S, Kingston J et al. Protein components specifically associated with prespliceosome and spliceosome complexes. Genes Dev 1992; 6(10):1986-2000.
8. Jamison SF, Crow A, Garcia-Blanco MA. The spliceosome assembly pathway in mammalian extracts. Mol Cell Biol 1992; 12(10):4279-4287.
9. Michaud S, Reed R. An ATP-independent complex commits pre-mRNA to the mammalian spliceosome assembly pathway. Genes Dev 1991; 5(12B):2534-2546.
10. Ruskin B, Zamore PD, Green MR. A factor, U2AF, is required for U2 snRNP binding and splicing complex assembly. Cell 1988; 52(2):207-219.
11. Zamore PD, Patton JG, Green MR. Cloning and domain structure of the mammalian splicing factor U2AF. Nature 1992; 355(6361):609-614.
12. Lim SR, Hertel KJ. Commitment to splice site pairing coincides with A complex formation. Mol Cell 2004; 15(3):477-483.
13. Kotlajich MV, Crabb TL, Hertel KJ. Spliceosome assembly pathways for different types of alternative splicing converge during commitment to splice site pairing in the A complex. Mol Cell Biol 2009; 29(4):1072-1082.
14. Krainer AR, Maniatis T. Multiple factors including the small nuclear ribonucleoproteins U1 and U2 are necessary for pre-mRNA splicing in vitro. Cell 1985; 42(3):725-736.
15. Krainer AR, Conway GC, Kozak D. The essential pre-mRNA splicing factor SF2 influences 5′ splice site selection by activating proximal sites. Cell 1990; 62(1):35-42.

16. Ge H, Manley JL. A protein factor, ASF, controls cell-specific alternative splicing of SV40 early pre-mRNA in vitro. Cell 1990; 62(1):25-34.

17. Krainer AR, Conway GC, Kozak D. Purification and characterization of pre-mRNA splicing factor SF2 from HeLa cells. Genes Dev 1990; 4(7):1158-1171.

18. Fu XD, Maniatis T. The 35-kDa mammalian splicing factor SC35 mediates specific interactions between U1 and U2 small nuclear ribonucleoprotein particles at the 3' splice site. Proc Natl Acad Sci U S A 1992; 89(5):1725-1729.

19. Boggs RT, Gregor P, Idriss S et al. Regulation of sexual differentiation in D. melanogaster via alternative splicing of RNA from the transformer gene. Cell 1987; 50(5):739-747.

20. Amrein H, Gorman M, Nothiger R. The sex-determining gene tra-2 of Drosophila encodes a putative RNA binding protein. Cell 1988; 55(6):1025-1035.

21. Roth MB, Zahler AM, Stolk JA. A conserved family of nuclear phosphoproteins localized to sites of polymerase II transcription. J Cell Biol 1991; 115:587-596.

22. Zahler AM, Lane WS, Stolk JA et al. SR proteins: a conserved family of pre-mRNA splicing factors. Genes Dev 1992; 6:837-847.

23. Shepard PJ, Hertel KJ. The SR protein family. Genome Biol 2009; 10(10):242.

24. Graveley BR, Hertel KJ. SR proteins. Encyclopedia of Life Sciences. Chichester: John Wiley & Sons, Ltd; 2005.

25. Long JC, Caceres JF. The SR protein family of splicing factors: master regulators of gene expression. Biochem J 2009; 417(1):15-27.

26. Manley JL, Krainer AR. A rational nomenclature for serine/arginine-rich protein splicing factors (SR proteins). Genes Dev 24(11):1073-1074.

27. Graveley BR. Sorting out the complexity of SR protein functions. RNA 2000; 6(9):1197-1211.

28. Blencowe BJ, Issner R, Kim J et al. New proteins related to the Ser-Arg family of splicing factors. RNA 1995; 1(8):852-865.

29. Blencowe BJ, Bowman JA, McCracken S et al. SR-related proteins and the processing of messenger RNA precursors. Biochem Cell Biol 1999; 77(4):277-291.

30. Blencowe BJ, Issner R, Nickerson JA et al. A coactivator of pre-mRNA splicing. Genes Dev 1998; 12(7):996-1009.

31. Fu X-D. The superfamily of arginine/serine-rich splicing factors. RNA 1995; 1:663-680.

32. Boucher L, Ouzounis CA, Enright AJ et al. A genome-wide survey of RS domain proteins. RNA 2001; 7(12):1693-1701.

33. Lutzelberger M, Gross T, Kaufer NF. Srp2, an SR protein family member of fission yeast: in vivo characterization of its modular domains. Nucleic Acids Res 1999; 27(13):2618-2626.

34. Gross T, Richert K, Mierke C et al. Identification and characterization of srp1, a gene of fission yeast encoding a RNA binding domain and a RS domain typical of SR splicing factors. Nucleic Acids Res 1998; 26(2):505-511.

35. Kress TL, Krogan NJ, Guthrie C. A single SR-like protein, Npl3, promotes pre-mRNA splicing in budding yeast. Mol Cell 2008; 32(5):727-734.

36. Plass M, Agirre E, Reyes D et al. Co-evolution of the branch site and SR proteins in eukaryotes. Trends Genet 2008; 24(12):590-594.

37. Escobar AJ, Arenas AF, Gomez-Marin JE. Molecular evolution of serine/arginine splicing factors family (SR) by positive selection. In Silico Biol 2006; 6(4):347-350.

38. Wu JY, Maniatis T. Specific interactions between proteins implicated in splice site selection and regulated alternative splicing. Cell 1993; 75:1061-1070.

39. Lin Y-S, Carey M, Ptashne M et al. How different eukaryotic transcriptional activators can co-operate promiscuously. Nature 1990; 345:359-361.

40. Shen H, Kan JL, Green MR. Arginine-serine-rich domains bound at splicing enhancers contact the branchpoint to promote prespliceosome assembly. Mol Cell 2004; 13(3):367-376.

41. Shen H, Green MR. A pathway of sequential arginine-serine-rich domain-splicing signal interactions during mammalian spliceosome assembly. Mol Cell 2004; 16(3):363-373.

42. Kohtz JD, Jamison SF, Will CL et al. Protein-protein interactions and 5' splice site recognition in mammalian mRNA precursors. Nature 1994; 368:119-124.

43. Teigelkamp S, Mundt C, Achsel T et al. The human U5 snRNP-specific 100-kD protein is an RS domain-containing, putative RNA helicase with significant homology to the yeast splicing factor Prp28p. RNA 1997; 3(11):1313-1326.

44. Fetzer S, Lauber J, Will CL et al. The [U4/U6.U5] tri-snRNP-specific 27K protein is a novel SR protein that can be phosphorylated by the snRNP-associated protein kinase. RNA 1997; 3(4):344-355.

45. Makarova OV, Makarov EM, Luhrmann R. The 65 and 110 kDa SR-related proteins of the U4/U6.U5 tri-snRNP are essential for the assembly of mature spliceosomes. EMBO J 2001; 20(10):2553-2563.

46. Shen H, Green MR. RS domains contact splicing signals and promote splicing by a common mechanism in yeast through humans. Genes Dev 2006; 20(13):1755-1765.
47. Cáceres JF, Misteli T, Screaton GR et al. Role of the modular domains of SR proteins in subnuclear localization and alternative splicing specificity. J Cell Biol 1997; 138(2):225-238.
48. Kataoka N, Bachorik JL, Dreyfuss G. Transportin-SR, a nuclear import receptor for SR proteins. J Cell Biol 1999; 145(6):1145-1152.
49. Lai MC, Lin RI, Huang SY et al. A human importin-beta family protein, transportin-SR2, interacts with the phosphorylated RS domain of SR proteins. J Biol Chem 2000; 275(11):7950-7957.
50. Cavaloc Y, Popielarz M, Fuchs JP et al. Characterization and cloning of the human splicing factor 9G8: a novel 35 kDa factor of the serine/arginine protein family. EMBO J 1994; 13(11):2639-2649.
51. Liu HX, Zhang M, Krainer AR. Identification of functional exonic splicing enhancer motifs recognized by individual SR proteins. Genes Dev 1998; 12(13):1998-2012.
52. Liu HX, Chew SL, Cartegni L et al. Exonic splicing enhancer motif recognized by human SC35 under splicing conditions. Mol Cell Biol 2000; 20(3):1063-1071.
53. Krainer AR, Mayeda A, Kozak D et al. Functional expression of cloned human splicing factor SF2: homology to RNA-binding proteins, U1 70K, and Drosophila splicing regulators. Cell 1991; 66(2):383-394.
54. Zuo P, Manley JL. The human splicing factor ASF/SF2 can specifically recognize pre-mRNA 5' splice sites. Proc Natl Acad Sci U S A 1994; 91(8):3363-3367.
55. Michaud S, Reed R. A functional association between the 5' and 3' splice site is established in the earliest prespliceosome complex (E) in mammals. Genes Dev 1993; 7(6):1008-1020.
56. Tacke R, Manley JL. The human splicing factors ASF/SF2 and SC35 possess distinct, functionally significant RNA binding specificities. EMBO J 1995; 14(14):3540-3551.
57. Cooper TA, Ordahl CP. Nucleotide substitutions within the cardiac troponin T alternative exon disrupt pre-mRNA alternative splicing. Nucleic Acids Res 1989; 17(19):7905-7921.
58. Streuli M, Saito H. Regulation of tissue-specific alternative splicing: exon-specific cis-elements govern the splicing of leukocyte common antigen pre-mRNA. EMBO J 1989; 8(3):787-796.
59. Sun Q, Mayeda A, Hampson RK et al. General splicing factor SF2/ASF promotes alternative splicing by binding to an exonic splicing enhancer. Genes Dev 1993; 7(12B):2598-2608.
60. Tian M, Maniatis T. A splicing enhancer complex controls alternative splicing of doublesex pre-mRNA. Cell 1993; 74:105-114.
61. Tian M, Maniatis T. A splicing enhancer exhibits both constitutive and regulated activities. Genes Dev 1994; 8:1703-1712.
62. Lavigueur A, La Branche H, Kornblihtt AR et al. A splicing enhancer in the human fibronectin alternate ED1 exon interacts with SR proteins and stimulates U2 snRNP binding. Genes Dev 1993; 7:2405-2417.
63. Ramchatesingh J, Zahler AM, Neugebauer KM et al. A subset of SR proteins activates splicing of the cardiac troponin T alternative exon by direct interactions with an exonic enhancer. Mol Cell Biol 1995; 15(9):4898-4907.
64. Gontarek RR, Derse D. Interactions among SR proteins, an exonic splicing enhancer, and a lentivirus Rev protein regulate alternative splicing. Mol Cell Biol 1996; 16(5):2325-2331.
65. Tian H, Kole R. Strong RNA splicing enhancers identified by a modified method of cycled selection interact with SR protein. J Biol Chem 2001; 276(36):33833-33839.
66. Sapra AK, Anko ML, Grishina I et al. SR protein family members display diverse activities in the formation of nascent and mature mRNPs in vivo. Mol Cell 2009; 34(2):179-190.
67. van Santen VL, Spritz RA. Alternative splicing of SV40 early pre-mRNA in vitro. Nucleic Acids Res 1986; 14(24):9911-9926.
68. Theissen H, Etzerodt M, Reuter R et al. Cloning of the human cDNA for the U1 RNA-associated 70K protein. EMBO J 1986; 5(12):3209-3217.
69. Lang KM, Spritz RA. In vitro splicing pathways of pre-mRNAs containing multiple intervening sequences? Mol Cell Biol 1987; 7(10):3428-3437.
70. Query CC, Bentley RC, Keene JD. A common RNA recognition motif identified within a defined U1 RNA binding domain of the 70K U1 snRNP protein. Cell 1989; 57(1):89-101.
71. Smith V, Barrell BG. Cloning of a yeast U1 snRNP 70K protein homologue: functional conservation of an RNA-binding domain between humans and yeast. EMBO J 1991; 10(9):2627-2634.
72. Query CC, Bentley RC, Keene JD. A specific 31-nucleotide domain of U1 RNA directly interacts with the 70K small nuclear ribonucleoprotein component. Mol Cell Biol 1989; 9(11):4872-4881.
73. Oubridge C, Krummel DA, Leung AK et al. Interpreting a low resolution map of human U1 snRNP using anomalous scatterers. Structure 2009; 17(7):930-938.
74. Pomeranz Krummel DA, Oubridge C, Leung AK et al. Crystal structure of human spliceosomal U1 snRNP at 5.5 A resolution. Nature 2009; 458(7237):475-480.
75. Zuo P, Maniatis T. The splicing factor U2AF^{35} mediates critical protein-protein interactions in constitutive and enhancer-dependent splicing. Genes Dev 1996; 10:1356-1368.

76. Rudner DZ, Kanaar R, Breger KS et al. Interaction between subunits of heterodimeric splicing factor U2AF is essential in vivo. Mol Cell Biol 1998; 18(4):1765-1773.
77. Kan JL, Green MR. Pre-mRNA splicing of IgM exons M1 and M2 is directed by a juxtaposed splicing enhancer and inhibitor. Genes Dev 1999; 13(4):462-471.
78. Lynch KW, Maniatis T. Synergistic interactions between two distinct elements of a regulated splicing enhancer. Genes Dev 1995; 9:284-293.
79. Zhou P, Lugovskoy AA, Wagner G. A solubility-enhancement tag (SET) for NMR studies of poorly behaving proteins. J Biomol NMR 2001; 20(1):11-14.
80. Hargous Y, Hautbergue GM, Tintaru AM et al. Molecular basis of RNA recognition and TAP binding by the SR proteins SRp20 and 9G8. EMBO J 2006; 25(21):5126-5137.
81. Cavaloc Y, Bourgeois CF, Kister L et al. The splicing factors 9G8 and SRp20 transactivate splicing through different and specific enhancers. RNA 1999; 5(3):468-483.
82. Schaal TD, Maniatis T. Selection and characterization of pre-mRNA splicing enhancers: identification of novel SR protein-specific enhancer sequences. Mol Cell Biol 1999; 19(3):1705-1719.
83. Ngo JC, Chakrabarti S, Ding JH et al. Interplay between SRPK and Clk/Sty kinases in phosphorylation of the splicing factor ASF/SF2 is regulated by a docking motif in ASF/SF2. Mol Cell 2005; 20(1):77-89.
84. Ngo JC, Giang K, Chakrabarti S et al. A sliding docking interaction is essential for sequential and processive phosphorylation of an SR protein by SRPK1. Mol Cell 2008; 29(5):563-576.
85. Ma CT, Hagopian JC, Ghosh G et al. Regiospecific phosphorylation control of the SR protein ASF/SF2 by SRPK1. J Mol Biol 2009; 390(4):618-634.
86. Hagopian JC, Ma CT, Meade BR et al. Adaptable molecular interactions guide phosphorylation of the SR protein ASF/SF2 by SRPK1. J Mol Biol 2008; 382(4):894-909.
87. Tian M. Positive regulation of pre-mRNA splicing [PhD]. Biochemistry and Molecular Biology. Cambridge: Harvard University; 1993.
88. Blencowe BJ. Exonic splicing enhancers: mechanism of action, diversity and role in human genetic diseases. Trends Biochem Sci 2000; 25(3):106-110.
89. Schaal TD, Maniatis T. Multiple distinct splicing enhancers in the protein-coding sequences of a constitutively spliced pre-mRNA. Mol Cell Biol 1999; 19(1):261-273.
90. Fairbrother WG, Yeh RF, Sharp PA et al. Predictive identification of exonic splicing enhancers in human genes. Science 2002; 297(5583):1007-1013.
91. Tuerk C, Gold L. Systematic evolution of ligands by exponential enrichment: RNA ligands to bacteriophage T4 DNA polymerase. Science 1990; 249(4968):505-510.
92. Shi H, Hoffman BE, Lis JT. A specific RNA hairpin loop structure binds the RNA recognition motifs of the Drosophila SR protein B52. Mol Cell Biol 1997; 17(5):2649-2657.
93. Tacke R, Tohyama M, Ogawa S et al. Human Tra2 proteins are sequence-specific activators of pre-mRNA splicing. Cell 1998; 93:139-148.
94. Zhang XH, Chasin LA. Computational definition of sequence motifs governing constitutive exon splicing. Genes Dev 2004; 18(11):1241-1250.
95. Cartegni L, Wang J, Zhu Z et al. ESEfinder: a web resource to identify exonic splicing enhancers. Nucleic Acids Res 2003; 31(13):3568-3571.
96. Mayeda A, Zahler AM, Krainer AR et al. Two members of a conserved family of nuclear phosphoproteins are involved in pre-mRNA splicing. Proc Natl Acad Sci U S A 1992; 89(4):1301-1304.
97. Fu XD, Maniatis T. Factor required for mammalian spliceosome assembly is localized to discrete regions in the nucleus. Nature 1990; 343(6257):437-441.
98. Fu X-D. Specific commitment of different pre-mRNAs to splicing by single SR proteins. Nature 1993; 365:82-85.
99. Zhang W-J, Wu JY. Functional properties of p54, a novel SR protein active in constitutive and alternative splicing. Mol Cell Biol 1996; 16(10):5400-5408.
100. Staknis D, Reed R. SR proteins promote the first specific recognition of pre-mRNA and are present together with U1 snRNP in a general splicing enhancer complex. Mol Cell Biol 1994; 14:7670-7682.
101. Crispino JD, Blencowe BJ, Sharp PA. Complementation by SR proteins of pre-mRNA splicing reactions depleted of U1 snRNP. Science 1994; 265(5180):1866-1869.
102. Tarn WY, Steitz JA. SR proteins can compensate for the loss of U1 snRNP functions in vitro. Genes Dev 1994; 8(22):2704-2717.
103. Boukis LA, Liu N, Furuyama S et al. Ser/Arg-rich protein-mediated communication between U1 and U2 small nuclear ribonucleoprotein particles. J Biol Chem 2004; 279(28):29647-29653.
104. Kennedy CF, Kramer A, Berget SM. A role for SRp54 during intron bridging of small introns with pyrimidine tracts upstream of the branch point. Mol Cell Biol 1998; 18(9):5425-5434.
105. Chiara MD, Reed R. A two-step mechanism for 5′ and 3′ splice-site pairing. Nature 1995; 375(6531):510-513.

106. Bruzik JP, Maniatis T. Enhancer-dependent interaction between 5' and 3' splice sites in trans. Proc Natl Acad Sci U S A 1995; 92(15):7056-7059.
107. Sanford JR, Bruzik JP. SR proteins are required for nematode trans-splicing in vitro. RNA 1999; 5(7):918-928.
108. Schneider M, Hsiao HH, Will CL et al. Human PRP4 kinase is required for stable tri-snRNP association during spliceosomal B complex formation. Nat Struct Mol Biol 2010; 17(2):216-221.
109. Schneider M, Will CL, Anokhina M et al. Exon definition complexes contain the tri-snRNP and can be directly converted into B-like precatalytic splicing complexes. Mol Cell 2010; 38(2):223-235.
110. Pozzoli U, Sironi M. Silencers regulate both constitutive and alternative splicing events in mammals. Cell Mol Life Sci 2005; 62(14):1579-1604.
111. Abdul-Manan N, Williams KR. hnRNP A1 binds promiscuously to oligoribonucleotides: utilization of random and homo-oligonucleotides to discriminate sequence from base-specific binding. Nucleic Acids Res 1996; 24(20):4063-4070.
112. Abdul-Manan N, O'Malley SM, Williams KR. Origins of binding specificity of the A1 heterogeneous nuclear ribonucleoprotein. Biochemistry 1996; 35(11):3545-3554.
113. Zhu J, Mayeda A, Krainer AR. Exon identity established through differential antagonism between exonic splicing silencer-bound hnRNP A1 and enhancer-bound SR proteins. Mol Cell 2001; 8(6):1351-1361.
114. Zuo P, Manley JL. Functional domains of the human splicing factor ASF/SF2. EMBO J 1993; 12(12):4727-4737.
115. Chandler SD, Mayeda A, Yeakley JM et al. RNA splicing specificity determined by the co-ordinated action of RNA recognition motifs in SR proteins. Proc Natl Acad Sci U S A 1997; 94(8):3596-3601.
116. Caceres JF, Krainer AR. Functional analysis of pre-mRNA splicing factor SF2/ASF structural domains. EMBO J 1993; 12(12):4715-4726.
117. Hicks MJ, Mueller WF, Shepard PJ et al. Competing upstream 5' splice sites enhance the rate of proximal splicing. Mol Cell Biol 2010; 30(8):1878-1886.
118. Hertel KJ, Lynch KW, Maniatis T. Common themes in the function of transcription and splicing enhancers. Curr Opin Cell Biol 1997; 9:350-357.
119. Lam BJ, Hertel KJ. A general role for splicing enhancers in exon definition. RNA 2002; 8(10):1233-1241.
120. Graveley BR, Hertel KJ, Maniatis T. A systematic analysis of the factors that determine the strength of pre-mRNA splicing enhancers. EMBO J 1998; 17(22):6747-6756.
121. Reed R, Maniatis T. A role for exon sequences and splice-site proximity in splice-site selection. Cell 1986; 46(5):681-690.
122. Mayeda A, Krainer AR. Regulation of alternative pre-mRNA splicing by hnRNP A1 and splicing factor SF2. Cell 1992; 68(2):365-375.
123. Fu XD, Mayeda A, Maniatis T et al. General splicing factors SF2 and SC35 have equivalent activities in vitro and both affect alternative 5' and 3' splice site selection. Proc Natl Acad Sci U S A 1992; 89(23):11224-11228.
124. Ibrahim EC, Schaal TD, Hertel KJ et al. Serine/arginine-rich protein-dependent suppression of exon skipping by exonic splicing enhancers. Proc Natl Acad Sci U S A 2005; 102(14):5002-5007.
125. Wang Z, Xiao X, Van Nostrand E et al. General and specific functions of exonic splicing silencers in splicing control. Mol Cell 2006; 23(1):61-70.
126. Zhang XH, Arias MA, Ke S et al. Splicing of designer exons reveals unexpected complexity in pre-mRNA splicing. RNA 2009; 15(3):367-376.
127. Yu Y, Maroney PA, Denker JA et al. Dynamic regulation of alternative splicing by silencers that modulate 5' splice site competition. Cell 2008; 135(7):1224-1236.
128. Ram O, Schwartz S, Ast G. Multifactorial interplay controls the splicing profile of Alu-derived exons. Mol Cell Biol 2008; 28(10):3513-3525.
129. Zahler AM, Neugebauer KM, Lane WS et al. Distinct functions of SR proteins in alternative pre-mRNA splicing. Science 1993; 260:219-222.
130. Zahler AM, Neugebauer KM, Stolk JA et al. Human SR proteins and isolation of a cDNA encoding SRp75. Mol Cell Biol 1993; 13(7):4023-4028.
131. Roscigno RF, Garcia-Blanco MA. SR proteins escort the U4/U6.U5 tri-snRNP to the spliceosome. RNA 1995; 1(7):692-706.
132. Hertel KJ, Maniatis T. Serine-arginine (SR)-rich splicing factors have an exon-independent function in pre-mRNA splicing. Proc Natl Acad Sci U S A 1999; 96(6):2651-2655.
133. Lou H, Neugebauer KM, Gagel RF et al. Regulation of alternative polyadenylation by U1 snRNPs and SRp20. Mol Cell Biol 1998; 18(9):4977-4985.
134. Kanopka A, Muhlemann O, Akusjarvi G. Inhibition by SR proteins of splicing of a regulated adenovirus pre-mRNA. Nature 1996; 381(6582):535-538.
135. Ibrahim EC, Schaal TD, Hertel KJ et al. Serine/arginine-rich protein-dependent suppression of exon skipping by exonic splicing enhancers. Proc Natl Acad Sci U S A 2005; 102(14):5002-5007.

136. Jumaa H, Nielsen PJ. The splicing factor SRp20 modifies splicing of its own mRNA and ASF/SF2 antagonizes this regulation. EMBO J 1997; 16(16):5077-5085.
137. ten Dam GB, Zilch CF, Wallace D et al. Regulation of alternative splicing of CD45 by antagonistic effects of SR protein splicing factors. J Immunol 2000; 164(10):5287-5295.
138. Gallego ME, Gattoni R, Stevenin J et al. The SR splicing factors ASF/SF2 and SC35 have antagonistic effects on intronic enhancer-dependent splicing of the beta-tropomyosin alternative exon 6A. EMBO J 1997; 16(7):1772-1784.
139. Soret J, Gattoni R, Guyon C et al. Characterization of SRp46, a novel human SR splicing factor encoded by a PR264/SC35 retropseudogene. Mol Cell Biol 1998; 18(8):4924-4934.
140. Simard MJ, Chabot B. SRp30c is a repressor of 3′ splice site utilization. Mol Cell Biol 2002; 22(12):4001-4010.
141. Wu JY, Kar A, Kuo D et al. SRp54 (SFRS11), a regulator for tau exon 10 alternative splicing identified by an expression cloning strategy. Mol Cell Biol 2006; 26(18):6739-6747.
142. Cowper AE, Caceres JF, Mayeda A et al. Serine-arginine (SR) protein-like factors that antagonize authentic SR proteins and regulate alternative splicing. J Biol Chem 2001; 276(52):48908-48914.
143. Shin C, Kleiman FE, Manley JL. Multiple properties of the splicing repressor SRp38 distinguish it from typical SR proteins. Mol Cell Biol 2005; 25(18):8334-8343.
144. Shin C, Feng Y, Manley JL. Dephosphorylated SRp38 acts as a splicing repressor in response to heat shock. Nature 2004; 427(6974):553-558.
145. Shin C, Manley JL. The SR protein SRp38 represses splicing in M phase cells. Cell 2002; 111(3):407-417.
146. Li J, Barnard DC, Patton JG. A unique glutamic acid-lysine (EK) domain acts as a splicing inhibitor. J Biol Chem 2002; 277(42):39485-39492.
147. Barnard DC, Li J, Peng R et al. Regulation of alternative splicing by SRrp86 through coactivation and repression of specific SR proteins. RNA 2002; 8(4):526-533.
148. Li J, Hawkins IC, Harvey CD et al. Regulation of alternative splicing by SRrp86 and its interacting proteins. Mol Cell Biol 2003; 23(21):7437-7447.
149. Exline CM, Feng Z, Stoltzfus CM. Negative and positive mRNA splicing elements act competitively to regulate human immunodeficiency virus type 1 vif gene expression. J Virol 2008; 82(8):3921-3931.
150. Johnson KW, Smith KA. Molecular cloning of a novel human cdc2/CDC28-like protein kinase. J Biol Chem 1991; 266(6):3402-3407.
151. Sukegawa J, Blobel G. A putative mammalian RNA helicase with an arginine-serine-rich domain colocalizes with a splicing factor. J Biol Chem 1995; 270(26):15702-15706.
152. Zhang WJ, Wu JY. Sip1, a novel RS domain-containing protein essential for pre-mRNA splicing. Mol Cell Biol 1998; 18(2):676-684.
153. Zhang M, Zamore PD, Carmo-Fonseca M et al. Cloning and intracellular localization of the U2 small nuclear ribonucleoprotein auxiliary factor small subunit. Proc Natl Acad Sci U S A 1992; 89(18):8769-8773.
154. Kaida D, Berg MG, Younis I et al. U1 snRNP protects pre-mRNAs from premature cleavage and polyadenylation. Nature 468(7324):664-668.
155. Park JW, Parisky K, Celotto AM et al. Identification of alternative splicing regulators by RNA interference in Drosophila. Proc Natl Acad Sci U S A 2004; 101(45):15974-15979.
156. Sharma S, Falick AM, Black DL. Polypyrimidine tract binding protein blocks the 5′ splice site-dependent assembly of U2AF and the prespliceosomal E complex. Mol Cell 2005; 19(4):485-496.
157. Mollet I, Barbosa-Morais NL, Andrade J et al. Diversity of human U2AF splicing factors. FEBS J 2006; 273(21):4807-4816.
158. Tronchere H, Wang J, Fu X. A protein related to splicing factor U2AF35 that interacts with U2AF65 and SR proteins in splicing of pre-mRNA. Nature 1997; 388:397-400.
159. Lynch KW, Maniatis T. Assembly of specific SR protein complexes on distinct regulatory elements of the Drosophila doublesex splicing enhancer. Genes Dev 1996; 10(16):2089-2101.
160. Sarkissian M, Winne A, Lafyatis R. The mammalian homolog of suppressor-of-white-apricot regulates alternative mRNA splicing of CD45 exon 4 and fibronectin IIICS. J Biol Chem 1996; 271(49):31106-31114.
161. Kondo S, Yamamoto N, Murakami T et al. Tra2 beta, SF2/ASF and SRp30c modulate the function of an exonic splicing enhancer in exon 10 of tau pre-mRNA. Genes Cells 2004; 9(2):121-130.
162. Sumner CJ. Molecular mechanisms of spinal muscular atrophy. J Child Neurol 2007; 22(8):979-989.
163. Eldridge AG, Li Y, Sharp PA et al. The SRm160/300 splicing coactivator is required for exon-enhancer function. Proc Natl Acad Sci U S A 1999; 96(11):6125-6130.
164. Blencowe BJ, Ouzounis CA. The PWI motif: a new protein domain in splicing factors. Trends Biochem Sci 1999; 24(5):179-180.
165. McCracken S, Longman D, Marcon E et al. Proteomic analysis of SRm160-containing complexes reveals a conserved association with cohesin. J Biol Chem 2005; 280(51):42227-42236.

166. Colwill K, Pawson T, Andrews B et al. The Clk/Sty protein kinase phosphorylates SR splicing factors and regulates their intranuclear distribution. EMBO J 1996; 15(2):265-275.
167. Prasad J, Colwill K, Pawson T et al. The protein kinase Clk/Sty directly modulates SR protein activity: both hyper- and hypophosphorylation inhibit splicing [In Process Citation]. Mol Cell Biol 1999; 19(10):6991-7000.
168. Behrens SE, Luhrmann R. Immunoaffinity purification of a [U4/U6.U5] tri-snRNP from human cells. Genes Dev 1991; 5(8):1439-1452.
169. Spector DL. Macromolecular domains within the cell nucleus. Annu Rev Cell Biol 1993; 9:265-315.
170. Spector DL. Nuclear organization and gene expression. Exp Cell Res 1996; 229(2):189-197.
171. Manley JL, Tacke R. SR proteins and splicing control. Genes Dev 1996; 10:1569-1579.
172. Hedley ML, Amrein H, Maniatis T. An amino acid sequence motif sufficient for subnuclear localization of an arginine/serine-rich splicing factor. Proc Natl Acad Sci U S A 1995; 92(25):11524-11528.
173. Jimenez-Garcia LF, Spector DL. In vivo evidence that transcription and splicing are coordinated by a recruiting mechanism. Cell 1993; 73(1):47-59.
174. Misteli T, Caceres JF, Spector DL. The dynamics of a pre-mRNA splicing factor in living cells. Nature 1997; 387(6632):523-527.
175. Misteli T, Caceres JF, Clement JQ et al. Serine phosphorylation of SR proteins is required for their recruitment to sites of transcription In vivo [In Process Citation]. J Cell Biol 1998; 143(2):297-307.
176. Tacke R, Chen Y, Manley JL. Sequence-specific RNA binding by an SR protein requires RS domain phosphorylation: creation of an SRp40-specific splicing enhancer. Proc Natl Acad Sci U S A 1997; 94(4):1148-1153.
177. Chew SL, Liu HX, Mayeda A et al. Evidence for the function of an exonic splicing enhancer after the first catalytic step of pre-mRNA splicing. Proc Natl Acad Sci U S A 1999; 96(19):10655-10660.
178. Mermoud JE, Cohen P, Lamond AI. Ser/Thr-specific protein phosphatases are required for both catalytic steps of pre-mRNA splicing. Nucleic Acids Res 1992; 20(20):5263-5269.
179. Mermoud JE, Cohen PT, Lamond AI. Regulation of mammalian spliceosome assembly by a protein phosphorylation mechanism. EMBO J 1994; 13(23):5679-5688.
180. Xiao S-H, Manley JL. Phosphorylation of the ASF/SF2 RS domain affects both protein-protein interactions and is necessary for splicing. Genes Dev 1997; 11:334-344.
181. Xiao SH, Manley JL. Phosphorylation-dephosphorylation differentially affects activities of splicing factor ASF/SF2. EMBO J 1998; 17(21):6359-6367.
182. Cao W, Jamison SF, Garcia-Blanco MA. Both phosphorylation and dephosphorylation of ASF/SF2 are required for pre-mRNA splicing in vitro. RNA 1997; 3(12):1456-1467.
183. Mattaj IW. RNA processing. Splicing in space. Nature 1994; 372(6508):727-728.
184. Okamoto Y, Onogi H, Honda R et al. cdc2 kinase-mediated phosphorylation of splicing factor SF2/ASF. Biochem Biophys Res Commun 1998; 249(3):872-878.
185. Soret J, Tazi J. Phosphorylation-dependent control of the pre-mRNA splicing machinery. Prog Mol Subcell Biol 2003; 31:89-126.
186. Lin S, Xiao R, Sun P et al. Dephosphorylation-dependent sorting of SR splicing factors during mRNP maturation. Mol Cell 2005; 20(3):413-425.
187. Huang Y, Steitz JA. Splicing factors SRp20 and 9G8 promote the nucleocytoplasmic export of mRNA. Mol Cell 2001; 7(4):899-905.
188. Huang Y, Gattoni R, Stevenin J et al. SR splicing factors serve as adapter proteins for TAP-dependent mRNA export. Mol Cell 2003; 11(3):837-843.
189. Huang Y, Yario TA, Steitz JA. A molecular link between SR protein dephosphorylation and mRNA export. Proc Natl Acad Sci U S A 2004; 101(26):9666-9670.
190. Lai MC, Tarn WY. Hypophosphorylated ASF/SF2 binds TAP and is present in messenger ribonucleoproteins. J Biol Chem 2004; 279(30):31745-31749.
191. Ma CT, Ghosh G, Fu XD et al. Mechanism of dephosphorylation of the SR protein ASF/SF2 by protein phosphatase 1. J Mol Biol 2010; 403(3):386-404.
192. Koizumi J, Okamoto Y, Onogi H et al. The subcellular localization of SF2/ASF is regulated by direct interaction with SR protein kinases (SRPKs). J Biol Chem 1999; 274(16):11125-11131.
193. Wang HY, Lin W, Dyck JA et al. SRPK2: a differentially expressed SR protein-specific kinase involved in mediating the interaction and localization of pre-mRNA splicing factors in mammalian cells. J Cell Biol 1998; 140(4):737-750.
194. Ding JH, Zhong XY, Hagopian JC et al. Regulated cellular partitioning of SR protein-specific kinases in mammalian cells. Mol Biol Cell 2006; 17(2):876-885.
195. Shi Y, Manley JL. A complex signaling pathway regulates SRp38 phosphorylation and pre-mRNA splicing in response to heat shock. Mol Cell 2007; 28(1):79-90.
196. Blaustein M, Pelisch F, Coso OA et al. Mammary epithelial-mesenchymal interaction regulates fibronectin alternative splicing via phosphatidylinositol 3-kinase. J Biol Chem 2004; 279(20):21029-21037.

197. Patel NA, Kaneko S, Apostolatos HS et al. Molecular and genetic studies imply Akt-mediated signaling promotes protein kinase CbetaII alternative splicing via phosphorylation of serine/arginine-rich splicing factor SRp40. J Biol Chem 2005; 280(14):14302-14309.
198. van der Houven van Oordt W, Diaz-Meco MT, Lozano J et al. The MKK(3/6)-p38-signaling cascade alters the subcellular distribution of hnRNP A1 and modulates alternative splicing regulation. J Cell Biol 2000; 149(2):307-316.
199. Fedorov A, Saxonov S, Fedorova L et al. Comparison of intron-containing and intron-lacking human genes elucidates putative exonic splicing enhancers. Nucleic Acids Res 2001; 29(7):1464-1469.
200. Fairbrother WG, Yeo GW, Yeh R et al. RESCUE-ESE identifies candidate exonic splicing enhancers in vertebrate exons. Nucleic Acids Res 2004; 32(Web Server issue):W187-W190.
201. Zhang XH, Chasin LA. Computational definition of sequence motifs governing constitutive exon splicing. Genes Dev 2004; 18(11):1241-1250.
202. Majewski J, Ott J. Distribution and characterization of regulatory elements in the human genome. Genome Res 2002; 12(12):1827-1836.
203. Barash Y, Calarco JA, Gao W et al. Deciphering the splicing code. Nature 2010; 465(7294):53-59.
204. Sanford JR, Coutinho P, Hackett JA et al. Identification of nuclear and cytoplasmic mRNA targets for the shuttling protein SF2/ASF. PLoS One 2008; 3(10):e3369.

CHAPTER 3

The Functions of Glycine-Rich Regions in TDP-43, FUS and Related RNA-Binding Proteins

Boris Rogelj,[1,2] Katherine S. Godin,[3] Christopher E. Shaw[1] and Jernej Ule*[,3]

Abstract

Glycine-rich regions form intrinsically unstructured domains within RNA-binding proteins. Although they lack a defined structure when alone in solution, these domains can form more defined structural elements when interacting with other proteins or with RNA. TDP-43 and FUS are RNA binding proteins with glycine-rich domains that form abnormal aggregates in amyotrophic lateral sclerosis (ALS) and frontotemporal dementia (FTD). The vast majority of mutations in familial ALS occur within the glycine-rich domain of TDP-43 as do about a third of FUS mutations. This chapter will review the various functions of some of the best characterised glycine-rich domains in RNA-binding proteins. Furthermore, the chapter will discuss how these findings inform on the possible functions of glycine-rich domains in TDP-43 and FUS.

Introduction

TAR DNA binding protein (TDP-43, encoded by the *TARDBP* gene) and fused in sarcoma (FUS) are two RNA-binding proteins (RBPs) that have attracted significant interest in recent years due to their involvement in neurodegenerative disorders. Approximately 60% of frontotemporal lobar degeneration (FTLD) and 90% of amyotrophic lateral sclerosis (ALS) cases are characterized pathologically by the TDP-43 inclusions. In addition, mutations in *TARDBP* or *FUS* gene are associated with familial ALS, and more rarely, with frontotemporal dementia (FTD, a syndrome caused by FTLD).[1-5] These familial cases are also characterized by TDP-43 or FUS inclusions implicating protein aggregation in disease pathogenesis. TDP-43 and FUS are predominantly nuclear proteins but in the pathological state they form inclusion in the cytoplasm of neurons and glia, with a corresponding decrease in nuclear staining.[3,6,7] TDP-43 cleavage products found in the cytoplasmic inclusions have been shown to alter splicing of a TDP-43 regulated exon in minigene studies, indicating that they are capable of interfering with wild-type TDP-43 function.[8] However, it is not yet clear whether inclusions of TDP-43 or FUS have a toxic gain of function or if their toxicity is a result of sequestration of the proteins from the nucleus, disrupting their role in gene expression.

TDP-43 and FUS regulate multiple aspects of gene expression, including transcription, alternative splicing and mRNA stability.[9-38] Both proteins contain glycine-rich regions that are predicted to be highly unstructured. Such glycine-rich regions are common in RBPs, but their function is poorly understood. In addition to TDP-43 and FUS, we will discuss the functions of glycine-rich

[1]MRC Centre for Neurodegeneration Research, Institute of Psychiatry, King's College London, London, UK; [2]Department of Biotechnology, Jozef Stefan Institute, Ljubljana, Slovenia; [3]MRC Laboratory of Molecular Biology, Cambridge, UK.
*Corresponding Author: Jernej Ule—Email: jule@mrc-lmb.cam.ac.uk

RNA Binding Proteins, edited by Zdravko J. Lorković.
©2012 Landes Bioscience.

regions in Sex-lethal (Sxl), heterogeneous nuclear ribonucleoprotein A1 (hnRNP A1), Fragile-X mental retardation protein (FMRP, encoded by *FMR1* gene), small nuclear ribonucleoprotein D1 polypeptide (SmD1, also SNRPD1) and U2 small nuclear RNA auxiliary factor 1 (U2AF1, also U2AF35). The glycine-rich regions in these proteins are part of predicted unstructured domains that contain several sub-regions that are enriched in a few additional amino acids. We show that glycine-rich regions have diverse functions in different RBPs, and we speculate how these differences might relate to the additional amino acids that are enriched in the glycine-rich regions.

General Properties of Intrinsically Unstructured Proteins

Intrinsically unstructured proteins lack a unique structure, either entirely or in part, when alone in solution. Studies of these proteins have revealed some general properties of unstructured regions.[39] Unstructured regions have a strong amino acid compositional bias, rich in hydrophilic and charged residues but often lacking bulky hydrophobic residues. This prevents the formation of a hydrophobic core necessary for a stable three-dimensional fold. Moreover, the large number of side-chain charges present under physiological conditions destabilizes any compact state. Furthermore, unstructured regions are enriched in amino acids that are targets for diverse posttranslational modifications, which can contribute to their conformational variability. These regions often adopt different folds when interacting with different partners. For instance, when unstructured regions are involved in RNA interactions, they can adopt distinct folds upon binding the RNA target. In these cases, it is the RNA tertiary structure that generates scaffolds and a binding pocket for recognition and discrimination amongst minimal elements of protein architecture.[40,41]

Another feature of intrinsically unstructured proteins is that their expression levels are precisely regulated. This includes low stochasticity in transcription and translation, and precise regulation of transcript clearance and proteolytic degradation.[42] In this chapter, we will also discuss several cases of RBPs containing unstructured regions that autoregulate their own expression, most often by regulating alternative splicing of their own transcript. The precise regulation of intrinsically unstructured proteins, together with their conformational variability, allows them to participate in multiple biological processes and to facilitate combinatorial regulation.[39]

Splicing Functions of Glycine-Rich Regions

Examination of glycine-rich regions of individual RBPs shows that each contains a small number of additional amino acids that are found in the putative unstructured region. These additional amino acids enriched in the glycine-rich regions contain polar (hydrophilic) residues, such as arginine, asparagine, glutamine, serine and tyrosine. Thus, glycine is the only amino acid enriched in the putative unstructured regions that contains a nonpolar (hydrophobic) side chain. Glycine is the smallest amino acid, with a single hydrogen atom as its side chain, allowing it to fit into tight spaces. These properties might underlie the propensity of glycine-rich regions to promote homo- and heteromeric interactions to create RNP complexes.

The additional amino acids enriched in the glycine-rich regions are likely to influence the interactions of these regions with other proteins or RNA, as well as determine the possibilities for posttranslational modifications. It is only due to our lack of knowledge that these diverse regions are collectively referred to as glycine-rich regions. In this chapter, we will first discuss the proteins that have enrichment of asparagine, glutamine and serine in glycine-rich regions. As we will discuss below, these proteins contain only one or two functional RNA-binding domains, and the glycine-rich regions often facilitate homo- and heteromeric interactions. This leads to formation of higher-order RNP complexes that allow co-operative RNA binding. We will discuss how this co-operative RNA binding might serve to increase the combinatorial complexity of splicing regulation.

Sxl

Sex-lethal (Sxl) is a *Drosophila* RBP that contains an unstructured region on the N-terminus, referred to as a glycine-rich N-terminus, and two RRM domains that bind uridine tracts in target RNAs (Fig. 1A). Due to its crucial role in *Drosophila* sexual differentiation, Sxl was one

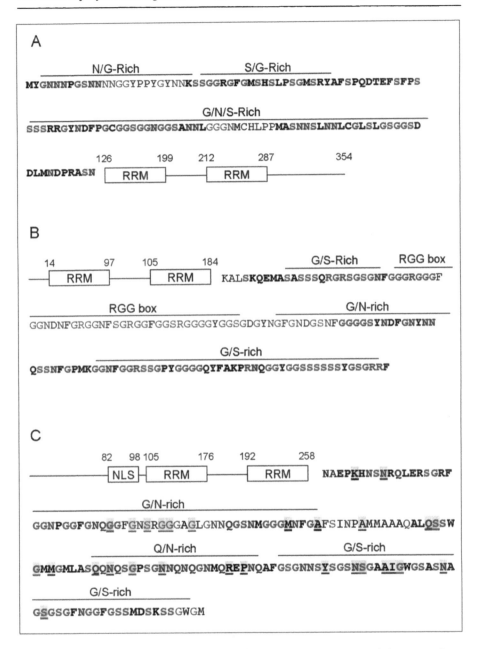

Figure 1. Domain structure and sequence of the glycine/serine/asparagine-rich domains in three RNA-binding proteins. A) *Drosophila* protein Sex-lethal (Sxl). B) Human protein heterogeneous nuclear ribonucleoprotein A1 (hnRNP A1). C) Human TAR DNA binding protein (TDP-43). The sequence of the protein region predicted to be disordered by Disopred2 software is shown, with highly disordered region in bold. Asparagines are labeled in blue, glutamines in purple, serines in pink, glycines in red and arginines in green. The amino acids in TDP-43 affected by disease-causing mutations are underlined and shaded in yellow.

of the first well-studied splicing regulators. As part of a cascade of genes that are regulated by sex-specific splicing, Sxl binds a uridine tract near the alternative 3' splice site of transformer (tra) pre-mRNA, thus blocking use of this site to give rise to the female-specific splicing pattern. This is achieved by competing with large subunit of the general splicing factor U2AF (*Drosophila* gene name is *U2af50*, and the human orthologue is U2AF2, also referred to as U2AF65) for binding to the poly-pyrimidine tract.[43] In addition, Sxl binds male-specific lethal-2 (msl-2) transcript to control its translation and thereby regulate dosage compensation.[44]

Sxl also autoregulates splicing of its own transcript by blocking inclusion of an alternative exon. Splicing autoregulation requires the intact N-terminal region of Sxl, even though this region is not required for translational regulation of msl-2 expression.[43,44] Although it has initially been proposed that the N-terminus is not necessary for tra regulation, later studies have proposed the opposite.[44,45] This discrepancy might have resulted from differences in how the groups have performed the experiments, which include different constructs (40-amino-acid deletion of the amino-terminal region in one case and 94-amino-acid deletion in the other). However, it appears likely that the Sxl N-terminal unstructured domain is required to regulate only a subset of its RNA targets.

Several observations point to distinct mechanisms in Sxl autoregulation that might involve the N-terminus. In contrast to the regulation of tra splicing, competition between Sxl and U2AF for binding to the uridine tract upstream of the exon in sxl transcript is insufficient to explain the autoregulation. This was shown by analysing effects of mutations within the poly(U) stretch upstream of the exon, and by expressing a fusion protein of U2AF and Sxl. Both had effects on splicing of tra, but not sxl transcripts. Instead, it was shown that Sxl needs to bind multiple uridine tracts upstream and downstream of the alternative exon in its own transcript in order to block its inclusion.[43] It was initially suggested that the N-terminal unstructured domain is implicated in homomeric protein-protein interactions that are required for Sxl multimerization on its own pre-mRNA.[46] However, later studies indicated that protein-protein interactions are mediated mainly by the RBDs.[47,48] Therefore, the role of the unstructured region in Sxl multimerisation remains to be fully resolved.

The Sxl N-terminus interacts with several other RBPs that and could mediate some of its splicing effects.[49] These include U2AF and sans fille (snf), a *Drosophila* orthologue of small nuclear ribonucleoprotein polypeptide A (SNRPA, also U1A).[50] U2AF plays a crucial role in 3' splice site recognition, whereas SNRPA is core component of the U1 small nuclear ribonucleoprotein particle (U1 snRNP), which recognises the 5' splice site. It was proposed that interaction of Sxl with U2AF and snf produces a complex that prevents the formation of a functional spliceosome (Fig. 2A).[43] Whereas the Sxl glycine-rich N-terminus was shown to be involved in interactions with glycine-rich regions in other proteins, such as Hrb87F (hnRNP A/B homologue) and hnRNP L, the interaction of Sxl protein with Snf occurs mainly through the RRM domain.[48,49]

Finally, the mechanism of Sxl autoregulation was also suggested to involve modulation of the second catalytic step of Sxl intron 2 splicing (Fig. 2B).[43] Interestingly, deletion of the whole unstructured region of Sxl (the N-terminal 94 amino acids, Fig. 1A) interfered with this function.[51] Moreover, the N-terminal region was shown to interact with splicing factor 45 (SPF45, also RBM17), a spliceosome component that functions in the second catalytic step of mRNA splicing.[51]

A detailed examination of the Sxl unstructured region shows that it is enriched in glycine (G), asparagine (N) and serine (S) (Fig. 1A). It contains sub-regions that are highly enriched in only one or two of these amino acids. For instance, the first 23 amino acids are enriched in asparagines and glycines, followed by 19 amino acids enriched in serines and glycines. It is likely that each of these sub-regions interacts with different proteins and, therefore, have different functions. However, as even the shortest deletion study of Sxl removed both the asparagine and serine-rich regions, more detailed Sxl deletion analyses would be required to fully understand the function of its unstructured region.[44]

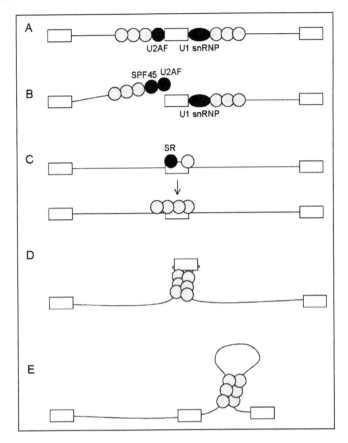

Figure 2. Splicing regulatory mechanisms that involve the glycine-rich regions of RNA-binding proteins. The studies of splicing functions of Sxl, hnRNP A1 and TDP-43 indicate that several splicing functions of these proteins reguire the glycine-rich unstructured regions. Exons are shown as boxes, introns as the lines connecting these boxes, and proteins as ovals. A) By interacting with *Drosophila* orthologues of U2AF2 (dark grey) and U1 snRNP (dark grey) at both sides of the exon, Sxl (light grey) was proposed to produce a complex that prevents the formation of a functional spliceosome. B) Sxl (light grey) was proposed to inhibit exon inclusion by interacting with a factor (SPF45, dark grey) that recognizes the 3' splice site in the second catalytic step of splicing (after cleavage of intron-exon junction). C) hnRNP A1 (light grey) can co-operatively spread on the pre-mRNA, which can act to unwind RNA hairpins and displace other proteins, such as SR proteins (dark grey), from the pre-mRNA. The TDP-43 crosslink enrichment at a distance from its high-affinity UG-rich binding sites indicates that it may have a similar ability for co-operative spreading. D,E) Formation of ribonucleoprotein (RNP) complexes (light grey) via homo- or heteromeric interactions between different RBPs can allow these RNPs to bind at multiple distal sites on the pre-mRNA and thereby change the RNA structure in a manner that can either silence (D) or enhance (E) exon inclusion.

hnRNP A1

hnRNP A1 is one of the most abundant nuclear RBPs. It contains two RRM domains at the N-terminus, which bind with high affinity to RNA sequences containing UAGGGA/U motif. The C-terminus contains a 124-amino-acid region that is predicted to be highly unstructured (Fig. 1B). This region contains domains rich in asparagine and glycine or serine and glycine, which resemble the Sxl N-terminus. In addition, it contains the RGG box, the function of which is discussed in the context of glycine and arginine-rich regions in a later section. Deletion of the C-terminal domain does not affect RNA binding, but it does interfere with autoregulation of hnRNP A1 alternative splicing.[52] Similar to Sxl, autoregulation requires binding to multiple distal RNA sites, and the C-terminal region is responsible for the ability of hnRNP A1 to silence splicing via co-operative RNA binding.[52-56]

Two models were proposed to explain how crosstalk between hnRNP A1 molecules bound at distal RNA sites affects splicing choice. The first proposed model is referred to as 'RNA looping model'. This model proposes that hnRNP A1 molecules bound at distal RNA sites can homomultimerise and thereby promote formation of an RNA loop.[52,57] The position of the two high-affinity binding sites determines the type of loop formed, which can allow hnRNP A1 to either silence or enhance exon inclusion (Fig. 2D,E).[52,57,58]

The second model is referred to as 'co-operative binding-dependent crosstalk' model. This model suggests that two distal high-affinity sites facilitate co-operative spreading of hnRNP A1 between the two sites, which can lead to displacement of other proteins and unwinding of RNA hairpins.[59] It was shown that initial hnRNP A1 binding to a high-affinity site in the human immunodeficiency virus type 1 (HIV-1) tat exon 3 is followed by co-operative spreading on the RNA. As hnRNP A1 propagates along the exon, it antagonizes binding of a serine/arginine-rich (SR) protein to an exonic splicing enhancer, thereby inhibiting splicing at that exon's alternative 3' splice site.[56] The C-terminal domain is required for the co-operative protein/protein interactions that underlie the co-operative spreading and was also proposed to be involved in protein/nucleic acid interactions.[53,54] Furthermore, hnRNP A1 was found to unwind an RNA hairpin upon binding, which required the C-terminal region.[59]

It is possible that the distance between the non-adjacent binding sites determines which of the two models explains the co-operative effects of hnRNP A1. If the two non-adjacent sites are positioned close to each other, then hnRNP A1 will be able to act via co-operative spreading that displaces RBPs bound to the intervening RNA region (Fig. 2C). However, if the two RNA binding sites are located a long distance from each other, hnRNP A1 might not be able to spread across the whole intervening RNA, but will therefore instead act via RNA looping.[57] Depending on the type of RNA loop that is promoted, exon inclusion could be either enhanced or silenced (Fig. 2D,E). Interestingly, polypyrimidine tract binding protein (PTB) can act via RNA looping without the need of multimerisation, which might be because it contains four RRM RNA-binding domains that enable its binding to multiple distal sites.[60,61] Therefore, one function of the unstructured domain in hnRNP A1 might be to allow it to form the multimeric complex that can bind to multiple distal sites. Furthermore, interactions between different RNA-binding proteins can also lead to multimeric complexes that bind to multiple distal sites to promote RNA looping, such as for instance interactions between hnRNP A/B and hnRNP F/H or between NOVA and FOX proteins.[58,62] The heteromeric interactions greatly enhance the combinatorial possibilities for regulation of RNA looping, and it is likely that the unstructured domains have an important contribution to these interactions.

TDP-43

TDP-43 has two N-terminal RRM domains, which bind UG repeat sequences with high affinity.[63] In addition, TDP-43 C-terminus is predicted to be unstructured, and has similar amino-acid enrichment as hnRNP A1, containing regions rich in glycine/serine or glycine/asparagine (Fig. 1C). Interestingly, the C-terminal domains of TDP-43 and hnRNP A1 interact with each other, and this interaction is required for TDP-43 to regulate alternative splicing.[64]

Mutations in the gene encoding TDP-43 (*TARDBP*) map to the C-terminal region and are associated with 1–4% of familial and sporadic ALS and have also been found in a few FTD cases (Fig. 1C).[5] In addition, TDP-43 pathology is evident in the majority of sporadic cases that lack any mutations in *TARDBP* gene. Given the similarity between TDP-43 and hnRNP A1 proteins, it is intriguing to speculate why TDP-43, but not hnRNP A1, has the propensity to form the toxic aggregates even in the absence of TDP-43 mutations. A comparison between the putative unstructured regions of the two proteins shows that TDP-43 contains a glutamine/asparagine (Q/N)-rich region, which is absent in hnRNP A1 (Fig. 1B,C). Interestingly, the Q/N-rich region in TDP-43 was required for its sequestration into detergent-insoluble inclusions formed by polyglutamine proteins.[65,66] Due to this result, it was proposed that the Q/N-rich region might have prion-like properties.[67]

TDP-43 has a propensity to bind RNA regions up to 100 nucleotides away from the UG repeats.[11] This indicated that the RNA binding properties of TDP-43 involve co-operative assembly, similar to hnRNP A1.[53,54,59] Since the C-terminus is required for such co-operative binding in hnRNP A1, we can speculate that the glycine-rich region in TDP-43 might play a similar role. The co-operative RNA binding is not a general property of RBPs, because another protein that also recognizes GU repeats, CUGBP, Elav-like family member 2 (CELF2), did not bind at positions further than 30 nucleotides from the GU repeat. Evidence for co-operative TDP-43 assembly was also seen in the vicinity of alternative exons that are silenced by TDP-43.[11] Such splicing silencing via multiple binding sites resembles the splicing autoregulation of Sxl and hnRNP A1, which requires multimerisation of these RBPs via the glycine-rich regions.[44]

TDP-43 autoregulates the levels of its own transcript by binding to multiple spaced sites on its own 3'UTR, indicating that co-operative assembly on its own RNA acts as a negative feedback loop to maintain the appropriate protein levels.[10,11,68] One aspect of autoregulation involves a splicing regulation of intron within the 3'UTR, which promotes degradation by nonsense-mediated mRNA decay (NMD).[10] A similar mode of autoregulation, where a splicing change leads to NMD, has been documented for PTB and most SR proteins, including autoregulation of an intron in the 3'UTR of SFRS1.[69-73] Unproductive splicing accounts for only part of the SFRS1 autoregulation, which occurs primarily at the translational level.[69] TDP-43 autoregulation via binding to the 3'UTR appears to similarly involve several levels of control, because one study showed that TDP-43 autoregulation is independent of NMD, and instead suggested that exosome-mediated degradation plays the primary role.[68]

Intrinsically unstructured proteins need an especially precise control of their expression levels.[42] Interestingly, some of the disease-causing mutations in TDP-43 were shown to increase the stability of TDP-43 protein, as well as increase its association with FUS protein.[74] Increased stability of TDP-43 could have detrimental effects, because increased expression levels of TDP-43 in yeast, *Drosophila* and chick are toxic.[75,76] Importantly, overexpression of TDP-43 is toxic only if its RNA binding function is intact.[76] One possible explanation for the requirement of RNA binding might be that increased levels of TDP-43 affect target RNAs in manners that are toxic for the cell, for instance by changing processing or translation of these RNAs. Alternatively, given that TDP-43 appears to be capable of co-operative assembly on RNA, increased expression of TDP-43 might have an increased tendency to multimerise upon RNA binding, which could initiate the formation of toxic aggregates.

Proteins with Glycine and Arginine-Rich Domains

Analysis of protein-RNA complexes showed that positively charged amino acid side chains of arginine and lysine contribute to about one-third of the protein-RNA hydrogen-bonds (see Chapter 9 by Cléry and Allain for details).[77] Therefore, arginine is a likely candidate to modulate interactions with RNA in domains rich in arginine and glycine, which are common in RNA-binding proteins.[78] Arginine in these domains is a common target for arginine methylation, which can affect protein-RNA interactions, protein localization and protein-protein interactions.[79] hnRNP proteins, such as hnRNP A1, are the primary substrates containing dimethylarginine in the cell nucleus.[80] Dimethylation of hnRNP proteins was shown to regulate their nuclear export.[81] The addition of methyl groups to amino acid side chains increases steric hindrance and removes amino hydrogens that normally participate in hydrogen bonding.[82] Thus, arginine and glycine-rich domains have a potential of interacting with RNA in a manner that would be regulated by arginine methylation.

RGG Box

The term RGG-box refers to clusters of closely spaced arginine-glycine-glycine tripeptides flanked by aromatic residues (Fig. 3A).[83] RGG boxes in FUS protein were shown to be unstructured when this protein is present free in solution.[19] RGG boxes are present in diverse RBPs, and were reported to have multiple functions. RGG-boxes in hnRNPK, hnRNPD, nucleolin and EWS were implicated in transcriptional regulation.[84-86] RGG boxes in hnRNP U, Ewing sarcoma breakpoint region 1 (EWSR1, also EWS), FUS and fragile X mental retardation 1 (FMR1, also FMRP) were reported to mediate interactions with RNA.[21,87-89] RGG-boxes are also able to mediate protein-protein interactions.[83] For instance, the RGG boxes in FUS interact with SR proteins.[22] RGG boxes in G3BP and nucleolin were suggested to have RNA/DNA helicase activity.[90] Finally, the RGG box in hnRNP A2 was reported to promote nuclear localization in a manner that is regulated by arginine methylation.[91]

The RNA-binding properties of the RGG box have been best characterized in FMRP, where in vitro RNA selection demonstrated that the FMRP RGG box binds intramolecular G-quartets.[88,92,93] G-quartets are nucleic acid structures in which four guanine residues are arranged in a planar conformation stabilized by Hoogsteen-type hydrogen bonds. The planar G-quartets stack on top of each other, giving rise to four-stranded helical structures.[94] (Fig. 3A). The formation of G-quartets is dependent on monovalent cations such as $K+$ and $Na+$, which is consistent with the observation that the FMRP RGG box bound its RNA target sites in a potassium-dependent manner.[88,92]

G-quartet formation can involve intermolecular RNA:RNA interactions.[95] Whereas intra-molecular folding of a G-rich strand containing four or more G-tracts is concentration independent, the intermolecular formation of G-quadruplex structure from two G-rich strands containing two G-tracts, or four strands each containing one G-track, is likely to require the aid of protein chaperones.[94] Similarly, nucleolin (which contains the RGG-domain) is required for stabilisation of G-quartet DNA and thereby repress transcription of *c-myc* gene.[96]

Intermolecular G-quartets could represent a flexible motif for mRNA oligomerization and RGG box binding to such G-quartets could stabilise RNA:RNA interactions. RGG boxes are strong targets for arginine methylation, as has been first shown in the case of hnRNP A1.[79,97] Therefore, interactions between G-quartets and RGG boxes could represent an important target for dynamic regulation of ribonucleoprotein complexes via posttranslational modifications of RBPs.

RG Domain

The second type of region enriched in arginine and glycine residues is a repeat of RG dinucleotide, which we will refer to as the RG domain. This domain is predicted to be highly unstructured (Fig. 3B). It is present in many RBPs, including four Sm/Lsm proteins that make up a core heptamer of the U snRNPs (SmB, SmD1, SmD3, and Lsm4). SnRNPs are essential components of the pre-mRNA splicing machinery, and their Sm cores are assembled in the cytoplasm.[98]

Assembly of the Sm cores onto the snRNAs is co-ordinated by survival of motor neuron protein (SMN) and a complex of several additional associated proteins. Interaction of the Tudor domain of SMN with RG domains of SmD1 and SmD3 (also named as SNRPD1 and SNRPD3) is crucial for its function in snRNP assembly.[99] Similar to the RGG box, the RG domains can also be modified by arginine methyltransferases. It was shown that symmetrical dimethylation of SmD1 and SmD3 is required for their high-affinity interaction with the SMN complex.[100] Coilin protein contains a similar RG domain that upon symmetrical dimethylation recruits SMN to Cajal bodies in cell nuclei. Furthermore, RA/RG motifs are also present in Piwi proteins. These proteins interact with a specific class of small noncoding RNAs, piwi-interacting RNAs (piRNAs) in germ cells. Symmetrical dimethylation of arginines within these motifs promotes their interaction with a specific set of Tudor domains.[101,102] These examples show that methyltransferases play an important role in regulating assembly of ribonucleoprotein complexes containing RBPs with the arginine- and glycine-rich regions.

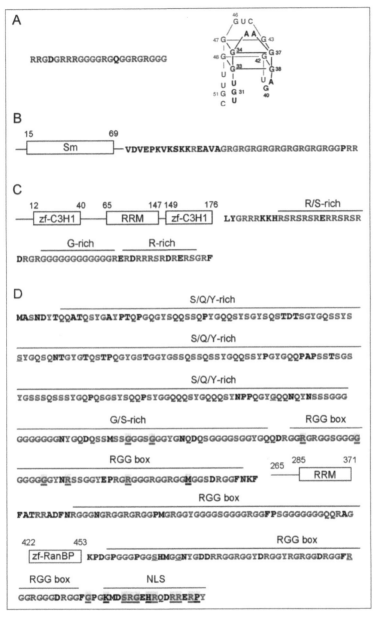

Figure 3. Glycine/arginine-rich domains in RNA-binding proteins. A) Sequence of the RGG box from fragile-X mental retardation protein (FMRP) and the structure of its RNA binding site, the G quartet. (Reprinted from: Darnell JC et al. Cell 2001; 107:489-499;[88] ©2001 with permission from Elsevier.) B) Domain structure and sequence of the small nuclear ribonucleoprotein D1 polypeptide (SmD1, also SNRPD1). C) Domain structure and sequence of the U2 small nuclear RNA auxiliary factor 1 (U2AF1, also U2AF35). D) Domain structure and sequence of the human fused in sarcoma (FUS). The sequence of the protein region predicted to be disordered by Disopred2 software is shown, with highly disordered region in bold. Tyrosines are labeled in blue, glutamines in purple, serines in pink, glycines in red and arginines in green. The amino acids in FUS affected by disease-causing mutations are underlined and shaded in yellow.

R-G Domain

The third type of region enriched in arginine and glycine residues is composed of consecutive glycines flanked by a proximal arginine-rich region. To differentiate from the RG domain discussed earlier, we will refer to this region as R-G domain. An example of such R-G domain is present in the predicted unstructured region of U2AF1 (Fig. 3C). In this protein, the R-G region is located next to the arginine and serine-rich (RS) domain, which is also predicted to be unstructured. Similar R-G domains are also present in several other important SR proteins (i.e., proteins that contain the RS domain), including splicing factor, arginine/serine-rich 1 (SFRS1, formerly ASF/SF2) and small nuclear ribonucleoprotein 70kDa (SNRP70, also U170K), as well as some hnRNP proteins, including heterogeneous nuclear ribonucleoprotein L (hnRNP L).

U2AF1, SFRS1 and SNRP70 are core splicing factors required for the early steps of splicing process. U2AF1 recognises the 3' splice site, SFRS1 the exonic splicing enhancers and SNRP70 is part of the U1 snRNP that recognises the 5' splice site. According to the exon definition model, interactions between the RS domains of proteins that recognize different regulatory elements on the exon promote recognition of exon as a unit by the splicing machinery.[103] Recognition of an exon as a unit was proposed to increase the fidelity of splicing.[104] Interestingly, the R-G domain in U2AF1 is more evolutionarily conserved compared to RS domain, as evident in the *C. elegans* orthologue uaf-2, which contains an extended R-G domain, but lacks the RS domain.[105] However, in contrast to the well-characterised functions of RS domains, little is known about the functions of R-G domains. It would therefore be interesting to explore the potential role of R-G domains in exon definition by promoting interactions between U2AF1, SFRS1 and SNRP70.

FUS

FUS, EWS and RNA polymerase II TATA box binding protein (TBP)-associated factor, 68kDa (TAF15) form a group of highly related FET (FUS, EWS, TAF15) family RBPs.[106] Their N terminus contains a glutamine-, serine-, and tyrosine-rich region that functions as a transcriptional activation domain when fused to a heterologous DNA binding domain.[107,108] This is followed by multiple domains involved in interactions with RNA: an RRM motif flanked by several RGG boxes and a C2C2 zinc finger. With the exception of the RRM and zinc finger domains, the remaining regions of FET proteins are predicted to be highly unstructured (Fig. 3D). FET proteins are involved in several aspects of gene expression, including transcription and splicing regulation and coupling of the transcriptional and splicing machinery.[3,109,110] FETs are targets for tyrosine and serine phosphorylation, as well as arginine methylation of the RGG boxes.[111-113] RGG boxes can self-associate and also play a role in nuclear import.[114] Similar to the association with RG-rich region in SmD1, the RGG box of EWS was also found to interact with SMN.[114]

The zinc finger domain belongs to the RanBP2-type and plays the predominant role in RNA recognition by recognising G-rich sequences.[19] The RNA specificity, if any, of the RGG boxes is not yet known. By analogy with the RGG box of FMRP, it would therefore be important to test if the RGG boxes of FET proteins also associate with G quartet RNA.

Majority of the ALS-associated FUS mutations fall within the PY-type nuclear localization signal present at the C-terminus of FUS.[2-5] They affect the nuclear import of FUS, bringing about cytoplasmic accumulation and in some instances formation of stress granules.[115-118]

Given the high similarity between FUS, EWS and TAF15 proteins, it is intriguing to speculate why FUS, but not other FET family RBPs, form toxic aggregates in neurons when mutated. A comparison between the putative unstructured regions of the three proteins shows that FUS contains a glycine/serine (G/S)-rich region, which is not present in other FET family proteins (Fig. 3D). This region is immediately preceded by an asparagine (Q)-rich region. This is similar to the C-terminus of TDP-43, which contains an N/Q-rich region followed by G/S-rich region (Fig. 1C). Due to this similarity, it was proposed that this domain might have prion-like properties.[67] Interestingly, about a third of the FUS mutations are found in the G/S-rich region, therefore it is of great importance to understand the function of this region of the protein.

Conclusion

The RNA-binding proteins discussed in this chapter indicate that glycine-rich regions can have diverse functions, which mainly depend on the few additional amino-acids that can be present in glycine-rich regions. Analysis of these amino acids indicates that the glycine-rich region in TDP-43 is most similar to proteins such as Sxl and hnRNP A1, whereas the glycine-rich region in FUS is most similar to proteins containing RGG boxes, such as FMR1. Studies of Sxl and hnRNP A1 indicate that their glycine-rich regions are required for multiple mechanisms of splicing regulation (Fig. 2). It is notable that similar to TDP-43, these regions are enriched in serines and asparagines. It remains to be seen if these additional amino acids contribute to the ability of these proteins for co-operative spreading on the target RNAs (Fig. 2C).

It is possible that glycine-rich regions are required for regulation of a specific set of RNA targets. As shown in the cases of Sxl and hnRNP A1, the unstructured region is particularly important for autoregulation, where RBPs need to co-operatively bind multiple sites on their own transcript. So far, however, only a small number of RNA targets of each protein have been evaluated. Therefore, a comprehensive exploration of RNA targets of a specific RBP will be required for better insight into the splicing functions of their glycine-rich regions. A large number of RNAs regulated by TDP-43 were recently identified, therefore it will be important to evaluate the role of its glycine-rich region in the regulation of these RNAs.[10,11]

Another poorly understood aspect of glycine-rich regions is the role of their posttranslational modifications. The best understood is the effect of arginine methylation in R/G-rich domains on protein-protein interactions. However, little is known about the effect of these modifications on protein-RNA interactions. Furthermore, even though G/S-rich regions are common in glycine-rich domains, the effect of serine phosphorylation of these regions has not yet been evaluated. Further experimental and computational studies of glycine-rich regions will be necessary to provide insights into these questions, and thereby contribute to our understanding of the effects of disease-causing mutations in TDP-43 and FUS proteins.

Acknowledgment

The authors would like to thank Julian Konig and Josh Witten for their instructive comments on this manuscript.

References

1. Sreedharan J, Blair IP, Tripathi VB et al. TDP-43 mutations in familial and sporadic amyotrophic lateral sclerosis. Science 2008; 319:1668-1672.
2. Vance C, Rogelj B, Hortobagyi T et al. Mutations in FUS, an RNA processing protein, cause familial amyotrophic lateral sclerosis type 6. Science 2009; 323:1208-1211.
3. Lagier-Tourenne C, Polymenidou M, Cleveland DW. TDP-43 and FUS/TLS: emerging roles in RNA processing and neurodegeneration. Hum Mol Genet 2010; 19:R46-R64.
4. Kwiatkowski TJ Jr, Bosco DA, Leclerc AL et al. Mutations in the FUS/TLS gene on chromosome 16 cause familial amyotrophic lateral sclerosis. Science 2009; 323:1205-1208.
5. Mackenzie IR, Rademakers R, Neumann M. TDP-43 and FUS in amyotrophic lateral sclerosis and frontotemporal dementia. Lancet Neurol 2010; 9:995-1007.
6. Mackenzie IR, Munoz DG, Kusaka H et al. Distinct pathological subtypes of FTLD-FUS. Acta Neuropathol 2011; 121:207-218.
7. Neumann M, Sampathu DM, Kwong LK et al. Ubiquitinated TDP-43 in frontotemporal lobar degeneration and amyotrophic lateral sclerosis. Science 2006; 314:130-133.
8. Igaz LM, Kwong LK, Chen-Plotkin A et al. Expression Of TDP-43 C-terminal fragments in vitro recapitulates pathological features of TDP-43 proteinopathies. J Biol Chem 2009.
9. Kuo PH, Doudeva LG, Wang YT et al. Structural insights into TDP-43 in nucleic-acid binding and domain interactions. Nucleic Acids Res 2009; 37:1799-1808.
10. Polymenidou M, Lagier-Tourenne C, Hutt KR et al. Long pre-mRNA depletion and RNA missplicing contribute to neuronal vulnerability from loss of TDP-43. Nat Neurosci 2011.
11. Tollervey JR, Curk T, Rogelj B et al. Characterizing the RNA targets and position-dependent splicing regulation by TDP-43. Nat Neurosci 2011.

12. Nguyen CD, Mansfield RE, Leung W et al. Characterization of a family of RanBP2-type zinc fingers that can recognize single-stranded RNA. J Mol Biol 2011.
13. Kim SH, Shanware N, Bowler MJ et al. ALS-associated proteins TDP-43 and FUS/TLS function in a common biochemical complex to coregulate HDAC6 mRNA. J Biol Chem 2010.
14. Selamat W, Jamari I, Wang Y et al. TLS interaction with NMDA R1 splice variant in retinal ganglion cell line RGC-5. Neurosci Lett 2009; 450:163-166.
15. Sugiura T, Sakurai K, Nagano Y. Intracellular characterization of DDX39, a novel growth-associated RNA helicase. Exp Cell Res 2007; 313:782-790.
16. Yoshimura A, Fujii R, Watanabe Y et al. Myosin-Va facilitates the accumulation of mRNA/protein complex in dendritic spines. Curr Biol 2006; 16:2345-2351.
17. Fujii R, Takumi T. TLS facilitates transport of mRNA encoding an actin-stabilizing protein to dendritic spines. J Cell Sci 2005; 118:5755-5765.
18. Belly A, Moreau-Gachelin F, Sadoul R et al. Delocalization of the multifunctional RNA splicing factor TLS/FUS in hippocampal neurones: exclusion from the nucleus and accumulation in dendritic granules and spine heads. Neurosci Lett 2005; 379:152-157.
19. Iko Y, Kodama TS, Kasai N et al. Domain architectures and characterization of an RNA-binding protein, TLS. J Biol Chem 2004; 279:44834-44840.
20. Delva L, Gallais I, Guillouf C et al. Multiple functional domains of the oncoproteins Spi-1/PU.1 and TLS are involved in their opposite splicing effects in erythroleukemic cells. Oncogene 2004; 23:4389-4399.
21. Lerga A, Hallier M, Delva L et al. Identification of an RNA binding specificity for the potential splicing factor TLS. J Biol Chem 2001; 276:6807-6816.
22. Yang L, Embree LJ, Tsai S et al. Oncoprotein TLS interacts with serine-arginine proteins involved in RNA splicing. J Biol Chem 1998; 273:27761-27764.
23. Zinszner H, Sok J, Immanuel D et al. TLS (FUS) binds RNA in vivo and engages in nucleo-cytoplasmic shuttling. J Cell Sci 1997; 110 (Pt 15):1741-1750.
24. Zinszner H, Immanuel D, Yin Y et al. A topogenic role for the oncogenic N-terminus of TLS: nucleolar localization when transcription is inhibited. Oncogene 1997; 14:451-461.
25. McDonald KK, Aulas A, Destroismaisons L et al. TAR DNA-binding protein 43 (TDP-43) regulates stress granule dynamics via differential regulation of G3BP and TIA-1. Hum Mol Genet 2011.
26. Sephton CF, Cenik C, Kucukural A et al. Identification of neuronal RNA targets of TDP-43-containing ribonucleoprotein complexes. J Biol Chem 2011; 286:1204-1215.
27. Liu-Yesucevitz L, Bilgutay A, Zhang YJ et al. Tar DNA binding protein-43 (TDP-43) associates with stress granules: analysis of cultured cells and pathological brain tissue. PLoS ONE 2010; 5:e13250.
28. Dreumont N, Hardy S, Behm-Ansmant I et al. Antagonistic factors control the unproductive splicing of SC35 terminal intron. Nucleic Acids Res 2010; 38:1353-1366.
29. Fiesel FC, Voigt A, Weber SS et al. Knockdown of transactive response DNA-binding protein (TDP-43) downregulates histone deacetylase 6. EMBO J 2010; 29:209-221.
30. Volkening K, Leystra-Lantz C, Yang W et al. Tar DNA binding protein of 43 kDa (TDP-43), 14-3-3 proteins and copper/zinc superoxide dismutase (SOD1) interact to modulate NFL mRNA stability. Implications for altered RNA processing in amyotrophic lateral sclerosis (ALS). Brain Res. 2009; 1305:168-182.
31. Strong MJ, Volkening K, Hammond R et al. TDP43 is a human low molecular weight neurofilament (hNFL) mRNA-binding protein. Mol Cell Neurosci 2007; 35:320-327.
32. Casafont I, Bengoechea R, Tapia O et al. TDP-43 localizes in mRNA transcription and processing sites in mammalian neurons. J Struct Biol 2009; 167:235-241.
33. Bose JK, Wang IF, Hung L et al. TDP-43 overexpression enhances exon 7 inclusion during the survival of motor neuron pre-mRNA splicing. J Biol Chem 2008; 283:28852-28859.
34. Ayala YM, Misteli T, Baralle FE. TDP-43 regulates retinoblastoma protein phosphorylation through the repression of cyclin-dependent kinase 6 expression. Proc Natl Acad Sci U S A 2008; 105:3785-3789.
35. Ayala YM, Pagani F, Baralle FE. TDP43 depletion rescues aberrant CFTR exon 9 skipping. FEBS Lett 2006; 580:1339-1344.
36. Ayala YM, Pantano S, D'Ambrogio A et al. Human, Drosophila and C. elegans TDP43: nucleic acid binding properties and splicing regulatory function. J Mol Biol 2005; 348:575-588.
37. Buratti E, Brindisi A, Pagani F et al. Nuclear factor TDP-43 binds to the polymorphic TG repeats in CFTR intron 8 and causes skipping of exon 9: a functional link with disease penetrance. Am J Hum Genet 2004; 74:1322-1325.
38. Buratti E, Dork T, Zuccato E et al. Nuclear factor TDP-43 and SR proteins promote in vitro and in vivo CFTR exon 9 skipping. EMBO J 2001; 20:1774-1784.
39. Gsponer J, Babu MM. The rules of disorder or why disorder rules. Prog Biophys Mol Biol 2009; 99:94-103.

40. Patel DJ. Adaptive recognition in RNA complexes with peptides and protein modules. Curr Opin Struct Biol 1999; 9:74-87.
41. Battiste JL, Mao H, Rao NS et al. Alpha helix-RNA major groove recognition in an HIV-1 rev peptide-RRE RNA complex. Science 1996; 273:1547-1551.
42. Gsponer J, Futschik ME, Teichmann SA et al. Tight regulation of unstructured proteins: from transcript synthesis to protein degradation. Science 2008; 322:1365-1368.
43. Penalva LO, Sanchez L. RNA binding protein sex-lethal (Sxl) and control of Drosophila sex determination and dosage compensation. Microbiol Mol Biol Rev 2003; 67:343-359, table.
44. Yanowitz JL, Deshpande G, Calhoun G et al. An N-terminal truncation uncouples the sex-transforming and dosage compensation functions of sex-lethal. Mol Cell Biol 1999; 19:3018-3028.
45. Granadino B, Penalva LO, Green MR et al. Distinct mechanisms of splicing regulation in vivo by the Drosophila protein Sex-lethal. Proc Natl Acad Sci U S A 1997; 94:7343-7348.
46. Wang J, Bell LR. The Sex-lethal amino terminus mediates co-operative interactions in RNA binding and is essential for splicing regulation. Genes Dev 1994; 8:2072-2085.
47. Sakashita E, Sakamoto H. Protein-RNA and protein-protein interactions of the Drosophila sex-lethal mediated by its RNA-binding domains. J Biochem 1996; 120:1028-1033.
48. Samuels M, Deshpande G, Schedl P. Activities of the Sex-lethal protein in RNA binding and protein:protein interactions. Nucleic Acids Res 1998; 26:2625-2637.
49. Wang J, Dong Z, Bell LR. Sex-lethal interactions with protein and RNA. Roles of glycine-rich and RNA binding domains. J Biol Chem 1997; 272:22227-22235.
50. Nagengast AA, Stitzinger SM, Tseng CH et al. Sex-lethal splicing autoregulation in vivo: interactions between SEX-LETHAL, the U1 snRNP and U2AF underlie male exon skipping. Development 2003; 130:463-471.
51. Lallena MJ, Chalmers KJ, Llamazares S et al. Splicing regulation at the second catalytic step by Sex-lethal involves 3' splice site recognition by SPF45. Cell 2002; 109:285-296.
52. Blanchette M, Chabot B. Modulation of exon skipping by high-affinity hnRNP A1-binding sites and by intron elements that repress splice site utilization. EMBO J 1999; 18:1939-1952.
53. Nadler SG, Merrill BM, Roberts WJ et al. Interactions of the A1 heterogeneous nuclear ribonucleo-protein and its proteolytic derivative, UP1, with RNA and DNA: evidence for multiple RNA binding domains and salt-dependent binding mode transitions. Biochemistry 1991; 30:2968-2976.
54. Cobianchi F, Karpel RL, Williams KR et al. Mammalian heterogeneous nuclear ribonucleoprotein complex protein A1. Large-scale overproduction in Escherichia coli and co-operative binding to single-stranded nucleic acids. J Biol Chem 1988; 263:1063-1071.
55. Mayeda A, Munroe SH, Caceres JF, Krainer AR. Function of conserved domains of hnRNP A1 and other hnRNP A/B proteins. EMBO J 1994; 13:5483-5495.
56. Zhu J, Mayeda A, Krainer AR. Exon identity established through differential antagonism between exonic splicing silencer-bound hnRNP A1 and enhancer-bound SR proteins. Mol Cell 2001; 8:1351-1361.
57. Nasim FU, Hutchison S, Cordeau M et al. High-affinity hnRNP A1 binding sites and duplex-forming inverted repeats have similar effects on 5' splice site selection in support of a common looping out and repression mechanism. RNA 2002; 8:1078-1089.
58. Martinez-Contreras R, Fisette JF, Nasim FU et al. Intronic binding sites for hnRNP A/B and hnRNP F/H proteins stimulate pre-mRNA splicing. PLoS Biol 2006; 4:e21.
59. Okunola HL, Krainer AR. Co-operative-binding and splicing-repressive properties of hnRNP A1. Mol Cell Biol 2009; 29:5620-5631.
60. Chan RC, Black DL. Conserved intron elements repress splicing of a neuron-specific c-src exon in vitro. Mol Cell Biol 1995; 15:6377-6385.
61. Amir-Ahmady B, Boutz PL, Markovtsov V et al. Exon repression by polypyrimidine tract binding protein. RNA 2005; 11:699-716.
62. Zhang C, Frias MA, Mele A et al. Integrative modeling defines the Nova splicing-regulatory network and its combinatorial controls. Science 2010; 329:439-443.
63. Buratti E, Baralle FE. Characterization and functional implications of the RNA binding properties of nuclear factor TDP-43, a novel splicing regulator of CFTR exon 9. J Biol Chem 2001; 276:36337-36343.
64. Buratti E, Brindisi A, Giombi M et al. TDP-43 binds heterogeneous nuclear ribonucleoprotein A/B through its C-terminal tail: an important region for the inhibition of cystic fibrosis transmembrane conductance regulator exon 9 splicing. J Biol Chem 2005; 280:37572-37584.
65. Fuentealba RA, Udan M, Bell S et al. Interaction with polyglutamine aggregates reveals a Q/N-rich domain in TDP-43. J Biol Chem 2010; 285:26304-26314.
66. Elden AC, Kim HJ, Hart MP et al. Ataxin-2 intermediate-length polyglutamine expansions are associated with increased risk for ALS. Nature 2010; 466:1069-1075.
67. Udan M, Baloh RH. Implications of the prion-related Q/N domains in TDP-43 and FUS. Prion 2011; 5.

68. Ayala YM, De CL, vendano-Vazquez SE et al. TDP-43 regulates its mRNA levels through a negative feedback loop. EMBO J 2011; 30:277-288.
69. Sun S, Zhang Z, Sinha R et al. SF2/ASF autoregulation involves multiple layers of posttranscriptional and translational control. Nat Struct Mol Biol 2010; 17:306-312.
70. Ni JZ, Grate L, Donohue JP et al. Ultraconserved elements are associated with homeostatic control of splicing regulators by alternative splicing and nonsense-mediated decay. Genes Dev 2007; 21:708-718.
71. Lareau LF, Inada M, Green RE et al. Unproductive splicing of SR genes associated with highly conserved and ultraconserved DNA elements. Nature 2007; 446:926-929.
72. Wollerton MC, Gooding C, Wagner EJ et al. Autoregulation of polypyrimidine tract binding protein by alternative splicing leading to nonsense-mediated decay. Mol Cell 2004; 13:91-100.
73. Jumaa H, Nielsen PJ. The splicing factor SRp20 modifies splicing of its own mRNA and ASF/SF2 antagonizes this regulation. EMBO J 1997; 16:5077-5085.
74. Ling SC, Albuquerque CP, Han JS et al. ALS-associated mutations in TDP-43 increase its stability and promote TDP-43 complexes with FUS/TLS. Proc Natl Acad Sci U S A 2010; 107:13318-13323.
75. Johnson BS, McCaffery JM, Lindquist S et al. A yeast TDP-43 proteinopathy model: Exploring the molecular determinants of TDP-43 aggregation and cellular toxicity. Proc Natl Acad Sci U S A 2008; 105:6439-6444.
76. Voigt A, Herholz D, Fiesel FC et al. TDP-43-mediated neuron loss in vivo requires RNA-binding activity. PLoS ONE 2010; 5:e12247.
77. Bahadur RP, Zacharias M, Janin J. Dissecting protein-RNA recognition sites. Nucleic Acids Res 2008; 36:2705-2716.
78. Godin KS, Varani G. How arginine-rich domains co-ordinate mRNA maturation events. RNA Biol 2007; 4:69-75.
79. Bedford MT. Arginine methylation at a glance. J Cell Sci 2007; 120:4243-4246.
80. Liu Q, Dreyfuss G. In vivo and in vitro arginine methylation of RNA-binding proteins. Mol Cell Biol 1995; 15:2800-2808.
81. Shen EC, Henry MF, Weiss VH et al. Arginine methylation facilitates the nuclear export of hnRNP proteins. Genes Dev 1998; 12:679-691.
82. McBride AE, Silver PA. State of the arg: protein methylation at arginine comes of age. Cell 2001; 106:5-8.
83. Burd CG, Dreyfuss G. Conserved structures and diversity of functions of RNA-binding proteins. Science 1994; 265:615-621.
84. Lee MH, Mori S, Raychaudhuri P. Trans-Activation by the hnRNP K protein involves an increase in RNA synthesis from the reporter genes. J Biol Chem 1996; 271:3420-3427.
85. Dempsey LA, Hanakahi LA, Maizels N. A specific isoform of hnRNP D interacts with DNA in the LR1 heterodimer: canonical RNA binding motifs in a sequence-specific duplex DNA binding protein. J Biol Chem 1998; 273:29224-29229.
86. Alex D, Lee KA. RGG-boxes of the EWS oncoprotein repress a range of transcriptional activation domains. Nucleic Acids Res 2005; 33:1323-1331.
87. Kiledjian M, Dreyfuss G. Primary structure and binding activity of the hnRNP U protein: binding RNA through RGG box. EMBO J 1992; 11:2655-2664.
88. Darnell JC, Jensen KB, Jin P et al. Fragile X mental retardation protein targets G quartet mRNAs important for neuronal function. Cell 2001; 107:489-499.
89. Ohno T, Ouchida M, Lee L et al. The EWS gene, involved in Ewing family of tumors, malignant melanoma of soft parts and desmoplastic small round cell tumors, codes for an RNA binding protein with novel regulatory domains. Oncogene 1994; 9:3087-3097.
90. Costa M, Ochem A, Staub A et al. Human DNA helicase VIII: a DNA and RNA helicase corresponding to the G3BP protein, an element of the ras transduction pathway. Nucleic Acids Res 1999; 27:817-821.
91. Nichols RC, Wang XW, Tang J et al. The RGG domain in hnRNP A2 affects subcellular localization. Exp Cell Res 2000; 256:522-532.
92. Schaeffer C, Bardoni B, Mandel JL et al. The fragile X mental retardation protein binds specifically to its mRNA via a purine quartet motif. EMBO J 2001; 20:4803-4813.
93. Bole M, Menon L, Mihailescu MR. Fragile X mental retardation protein recognition of G quadruplex structure per se is sufficient for high affinity binding to RNA. Mol Biosyst 2008; 4:1212-1219.
94. Lipps HJ, Rhodes D. G-quadruplex structures: in vivo evidence and function. Trends Cell Biol 2009; 19:414-422.
95. Ramos A, Hollingworth D, Pastore A. G-quartet-dependent recognition between the FMRP RGG box and RNA. RNA 2003; 9:1198-1207.
96. Gonzalez V, Guo K, Hurley L et al. Identification and characterization of nucleolin as a c-myc G-quadruplex-binding protein. J Biol Chem 2009; 284:23622-23635.

97. Kim S, Merrill BM, Rajpurohit R et al. Identification of N(G)-methylarginine residues in human heterogeneous RNP protein A1: Phe/Gly-Gly-Gly-Arg-Gly-Gly-Gly/Phe is a preferred recognition motif. Biochemistry 1997; 36:5185-5192.

98. Kambach C, Walke S, Nagai K. Structure and assembly of the spliceosomal small nuclear ribonucleoprotein particles. Curr Opin Struct Biol 1999; 9:222-230.

99. Paushkin S, Gubitz AK, Massenet S et al. The SMN complex, an assemblyosome of ribonucleoproteins. Curr Opin Cell Biol 2002; 14:305-312.

100. Brahms H, Meheus L, de B et al. Symmetrical dimethylation of arginine residues in spliceosomal Sm protein B/B' and the Sm-like protein LSm4, and their interaction with the SMN protein. RNA 2001; 7:1531-1542.

101. Vagin VV, Wohlschlegel J, Qu J et al. Proteomic analysis of murine Piwi proteins reveals a role for arginine methylation in specifying interaction with Tudor family members. Genes Dev 2009; 23:1749-1762.

102. Liu K, Chen C, Guo Y et al. Structural basis for recognition of arginine methylated Piwi proteins by the extended Tudor domain. Proc Natl Acad Sci U S A 2010; 107:18398-18403.

103. Kohtz JD, Jamison SF, Will CL et al. Protein-protein interactions and 5'-splice-site recognition in mammalian mRNA precursors. Nature 1994; 368:119-124.

104. Robberson BL, Cote GJ, Berget SM. Exon definition may facilitate splice site selection in RNAs with multiple exons. Mol Cell Biol 1990; 10:84-94.

105. Zorio DA, Blumenthal T. U2AF35 is encoded by an essential gene clustered in an operon with RRM/ cyclophilin in Caenorhabditis elegans. RNA 1999; 5:487-494.

106. Bertolotti A, Lutz Y, Heard DJ et al. hTAF(II)68, a novel RNA/ssDNA-binding protein with homology to the pro-oncoproteins TLS/FUS and EWS is associated with both TFIID and RNA polymerase II. EMBO J 1996; 15:5022-5031.

107. May WA, Lessnick SL, Braun BS et al. The Ewing's sarcoma EWS/FLI-1 fusion gene encodes a more potent transcriptional activator and is a more powerful transforming gene than FLI-1. Mol Cell Biol 1993; 13:7393-7398.

108. Zinszner H, Albalat R, Ron D. A novel effector domain from the RNA-binding protein TLS or EWS is required for oncogenic transformation by CHOP. Genes Dev 1994; 8:2513-2526.

109. Yang L, Chansky HA, Hickstein DD. EWS.Fli-1 fusion protein interacts with hyperphosphorylated RNA polymerase II and interferes with serine-arginine protein-mediated RNA splicing. J Biol Chem 2000; 275:37612-37618.

110. Knoop LL, Baker SJ. The splicing factor U1C represses EWS/FLI-mediated transactivation. J Biol Chem 2000; 275:24865-24871.

111. Felsch JS, Lane WS, Peralta EG. Tyrosine kinase Pyk2 mediates G-protein-coupled receptor regulation of the Ewing sarcoma RNA-binding protein EWS. Curr Biol 1999; 9:485-488.

112. Belyanskaya LL, Gehrig PM, Gehring H. Exposure on cell surface and extensive arginine methylation of ewing sarcoma (EWS) protein. J Biol Chem 2001; 276:18681-18687.

113. Olsen RJ, Hinrichs SH. Phosphorylation of the EWS IQ domain regulates transcriptional activity of the EWS/ATF1 and EWS/FLI1 fusion proteins. Oncogene 2001; 20:1756-1764.

114. Shaw DJ, Morse R, Todd AG et al. Identification of a self-association domain in the Ewing's sarcoma protein: a novel function for arginine-glycine-glycine rich motifs? J Biochem 2010; 147:885-893.

115. Ito D, Seki M, Tsunoda Y et al. Nuclear transport impairment of amyotrophic lateral sclerosis-linked mutations in FUS/TLS. Ann Neurol 2011; 69:152-162.

116. Bosco DA, Lemay N, Ko HK et al. Mutant FUS proteins that cause amyotrophic lateral sclerosis incorporate into stress granules. Hum Mol Genet 2010.

117. Gal J, Zhang J, Kwinter DM et al. Nuclear localization sequence of FUS and induction of stress granules by ALS mutants. Neurobiol Aging 2010.

118. Dormann D, Rodde R, Edbauer D et al. ALS-associated fused in sarcoma (FUS) mutations disrupt Transportin-mediated nuclear import. EMBO J 2010; 29:2841-2857.

Chapter 4

Subcellular RNA Localization and Translational Control:
Mechanisms and Biological Significance in the Vertebrate Nervous System

Alessandro Quattrone,[1] Ralf Dahm[2] and Paolo Macchi*[,1]

Abstract

Eukaryotic cells adopt different mechanisms to control gene expression: Transcriptional regulation, post-transcriptional control of mRNA translation and turnover, and post-translational regulation of proteins. Another mechanism, the localization of RNAs, is emerging as an important process to restrict certain messages and proteins to specific subcellular domains and thus spatially control the expression of genes within cells. Messenger RNA localization has been studied in many organisms and cell types and research over the last decade has shown that homologues of key components of the mRNA localization machinery are conserved from yeast to mammals. In mammalian neurons, local translation of mRNAs in dendrites as well as growth cones allows for de novo synapse formation, the morphological re-arrangement of dendritic spines and for the regulation of the efficacy in the strength, e.g., electrical properties of existing synapses.

There are various mechanisms by which RNA translation may be regulated during transport. RNA stability control, translational repression, regulatory molecule recruitment and transport control have been implicated as molecular processes by which different RNA binding proteins exert their biological function in different species.

In this chapter we will highlight the biological relevance of mRNA transport and localized translation in the mammalian nervous system, the molecular players—RNAs and proteins—involved in this process and we will discuss the current state of knowledge concerning these mechanisms.

Introduction

Nothing is more revealing than movement.
—Martha Graham

Neurons are among the cell types with the most complex morphology. A mature mammalian neuron typically has a complex dendritic tree as well as a highly arborized axon protruding from the cell body (soma). Axon and dendrites themselves are further subdivided into a shaft as well as pre and post-synaptic sites, respectively. Developing and maintaining this morphology

[1]Centre for Integrative Biology, University of Trento, Trento, Italy; [2]Department of Biology, University of Padova, Padova, Italy.
*Corresponding Author: Paolo Macchi—Email: macchi@science.unitn.it

RNA Binding Proteins, edited by Zdravko J. Lorković.

requires a very precise control of where and when specific proteins are incorporated. This can be achieved by controlling the relevant proteins themselves, e.g., via their localization, activity and turnover, or by restricting their synthesis to sites where they are needed. The latter is accomplished through localizing their respective messenger RNAs (mRNAs) and keeping them translationally inactive until appropriate signals lift this block and initiate the synthesis of protein. As such, the localization of mRNAs to specific subcellular sites is a powerful mechanism to make proteins available only at these specific sites within a cell.

Neurons use this mechanism at various stages of their development. An early example of the importance of localizing mRNAs to different parts of the cell is the asymmetric cell divisions of neuroblasts where the two daughter cells receive different sets of messages and hence follow different fates: One remaining a stem cell that self-renews, while the progeny of the other terminally differentiates. A second example is the outgrowth and pathfinding of neurites—prospective axons and dendrites—during later stages of development. Here, proteins synthesized from localized mRNAs are crucial in driving the outgrowth of the neurites' tips and in directing the migrating growth cones along gradients of chemoattractants and repellants that prepattern the developing nervous system. Finally, also during the last stages of development of the nervous system and in mature neurons, RNA localization plays a pivotal role in the establishment and modification of synaptic contacts between neurons. This process is essential not only to interconnect the neurons with each other, but also to allow these contacts to be altered in order to enable the brain to adjust to its environment. In this context, it is critical that individual synapses can be specifically modified in response to neuronal activity. Given that a single neuron can form thousands of synapses this is no trivial task. The recruitment and translational activation of localized mRNAs to synapses that have received signals inducing synaptic plasticity are thought to be important for ensuring this specificity.

The local synthesis of proteins from mRNAs is therefore crucially involved in three key processes that shape the developing nervous system: (i) the determination of cell fate during the early proliferation of neuronal precursors, (ii) the correct wiring of the nervous system when axons and dendrites develop, and (iii) the formation as well as, importantly, the plasticity of synapses that underlies learning and memory.

Localizing the mRNA, rather than the corresponding protein itself, could have a number of advantages. For one, as indicated above, synthesizing proteins locally may afford a higher spatial precision than if the proteins were to be targeted to specific subcellular regions. Secondly, producing protein locally may be more rapid than having to rely on mRNA synthesis in the nucleus—which in a neuron may be a significant distance away—and require less protein to be synthesized. Thirdly, producing proteins locally alleviates the need to control their activity before they are incorporated into their target structures. Finally, local synthesis may facilitate the cotranslational incorporation of such proteins into their target structures.

Localizing mRNAs also poses a number of challenges: The RNA needs to be recognized as one that is to be localized and packaged into a messenger ribonucleoprotein particle (RNP) that (i) transports it to its final destination, (ii) keeps the mRNA translationally silent during this transport and when anchored to its target region, and (iii) lifts the translational block when appropriate stimuli are received. Whereas the number of localized mRNAs that have been identified has significantly grown during the past decade, the molecular details underlying this process are still largely unknown. For instance, comparatively few putative localization elements (LEs; cis-acting elements) have been discovered and for fewer it has been shown that they are required for the proper subcellular localization of an mRNA. Similarly, although a number of RNA-binding proteins (RBPs) associated with localized mRNAs (trans-acting factors) have been identified, their precise roles and how they are regulated are only poorly characterized. Lastly, the signals that induce the switch from a translationally repressed transport RNP to one whose mRNAs are actively transcribed have only been elucidated in very few cases.

Localized mRNAs in Neurons

In 1964, David Bodian wrote the following words at the end of his manuscript:

> *In view of the fact that the only known nerve growth factor is of protein nature and is highly specific, one may speculate that selective establishment of synaptic contacts may be determined by specific proteins synthesized at the synaptic membrane of the receptive neuron.*[1]

Bodian was postulating that mRNAs are present in dendrites and/or in the proximity of post-synaptic sites (Fig. 1). His hypothesis turned out to be correct two decades later. In fact, the story concerning the discovery of mRNAs in dendrites only started in the early 1980s. Transmission electron microscopy analyses showed synapse-associated polyribosome complexes localized in close proximity of post-synaptic sites in dendritic shaft in the central nervous system.[2] At the same time, the mRNAs encoding for the microtubule associated protein 2 (MAP2) and for the α-subunit of calcium/calmodulin-dependent protein kinase II alpha (CaMKIIα), respectively, were the first mRNAs showing dendritic localization.[3,4] Interestingly, the localization patterns of these two transcripts within the dendritic compartment are different suggesting distinctive biological functions for these two messages and the respective proteins in dendrites. In contrast to the CaMKIIα mRNA, which is found in more distal processes, the *MAP2* transcripts are restricted to more proximal regions.[5] Moreover, mRNAs encoding different isoforms of *MAP2* have been found in the somato-dendritic compartments where they might control the stability of the dendritic cytoskeleton.[6] A specific, 640 nucleotide-long sequence, the dendritic targeting element (or DTE), which is located in the *MAP2* mRNA's 3'-untranslated region (3' UTR), is necessary and sufficient to localize a reporter mRNA in dendrites of hippocampal neurons.[7]

CaMKIIα is a serine/threonine kinase that is found in many tissues and highly expressed in neurons where it can represent up to 2% of total protein in some brain regions. In adult neurons, CaMKII consists of holoenzymes composed of α- and β subunits. Several groups contributed to dissecting the signals mediating the dendritic transport and translation of the *CaMKIIα* transcript in cultured neurons.[8,9] However, Mayford and coworkers created the first mouse model where the DTE of *CaMKIIα* was removed from the genome.[10] This mutant mouse is characterized by the expression of a fully functional and stable protein-coding region of CaMKIIα, whose mRNA is, however, restricted to the cell body in hippocampal neurons. Genomic ablation of the DTE resulted in a significant reduction of CaMKIIα protein at post-synaptic densities. At the functional level, the neurons showed a reduction in late-phase of long-term potentiation (LTP), impairments in spatial memory, associative fear conditioning and object recognition memory.[10] This paper thus confirmed, for the first time, that targeting elements within the mRNA 3'UTR are necessary and sufficient for dendritic localization of the *CaMKIIα* mRNA in vivo and provided new insights into the nature and biological significance of dendritic mRNA localisation.

With the advent of new imaging and molecular techniques (e.g., fluorescence microscopy, in situ hybridizations and PCR), the list of mRNAs identified in the neuronal dendrites has grown to hundreds of transcripts. Among these there are proteins with key functions in neurons, such as *Arc* mRNA, which codes for a protein required for the internalization of AMPA receptors during long-term depression (LTD);[11,12] the mRNAs for the brain-derived neurotrophic factor BDNF and its receptor TrkB;[13] the mRNA for the translation elongation factor EF1A;[14] mRNAs coding for Pumilio2, a protein involved in translational control and for the RBP HuD;[15,16] *protein kinase Mzeta* mRNA (PKMzeta, the new form of PKC);[17] the mRNA for the splicing regulatory protein Sam68;[18] the mRNA coding for the protein Shank 1 that functions as a molecular scaffold in the post-synaptic density (PSD);[19,20] the transcript coding for the peptide hormone and neurotransmitter/neuromodulator vasopressin[21,22] and *EFA6A*[23] mRNA coding a guanine nucleotide exchange factor that activates the ADP-ribosylation factor 6 (ARF6) that is involved in dendritic morphogenesis.[24] In general, these transcripts encode proteins involved in cytoskeletal re-arrangements, signal transduction, trascriptional and translational control, trans-membrane signaling and proteins which act as growth factors.[25-29] These transcripts and their biological roles in neurons have been discussed in several recent reviews.[23,27,30-32] Nevertheless, the expanding number of newly identified

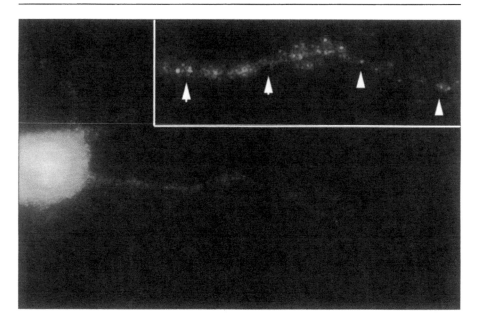

Figure 1. Localization of mRNAs in neurons in a cultured hippocampal neuron at 15 days in vitro (DIV). Fluorescent in situ hybridization (FISH) using a poly-dT anti sense probe has been used to detect poly-adenylated mRNAs. RNAs form discrete particles (indicated by arrowheads in the enlarged inset)both in the cell body and the dendritic compartment.

dendritically localized mRNAs increases the number of local functions that the corresponding proteins could exert in the neurons.

The localization of neuronal transcripts is not restricted to the dendritic compartment. In developing neurons, the transport of specific mRNAs to axonal growth cones and their subsequent local translation is crucial for the ability of neurons to respond to environmental cues.[33,34] Transcripts such as *β-actin* (Fig. 2), *cofilin* and *RhoA*, are locally translated in growth cones in response to extracellular guidance cues. This allows axonal survival and outgrowth in order to

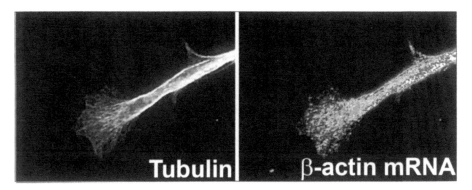

Figure 2. Subcellular localization of *β-actin* mRNA in cultured hippocampal neurons. High magnification of a neuronal growth cone at 4 DIV. Right panel: *β-actin* mRNA detected by FISH performed on a cultured mouse hippocampal neuron. RNPs-containing *β-actin* transcripts localize to the tip of the developing growth cone. Left panel: Immunostaining for tubulin protein. (Image courtesy of Gary Bassell and Kristy Welshhans.)

project towards the correct target area to form specific synapses. A short LE of 54 nucleotides, referred to as *zipcode*, has been identified in the 3'UTR of the *β-actin* mRNA. This sequence, which is absent in the other *actin* mRNA isoforms, is responsible for the distal targeting of the *β-actin* subunit to growth cones. The molecular process leading to *β-actin* mRNA transport and the mechanism of its subsequent local translation have recently been unraveled.[35-39] In contrast to mature dendrites and developing neurites, for decades no polysomes were detected in mature axons suggesting that RNA localization and subsequent translation do not occur in this compartment. However, this assumption has recently been challenged.[33] An axonal localization signal of 240 nucleotides was identified in the 3'UTR of *tau* mRNA, which has been detected in mature axons of PC19 cells.[40] Lately, the mRNA encoding the K-opioid receptor (KOR) was detected in mature axons of primary neurons of dorsal root ganglia by fluorescent in situ hybridization (FISH).[41] The authors showed the mRNA transport and the regulation of presynaptic translation of nonstructural proteins and a mechanism of depolarization-stimulated axonal mRNA redistribution for localized translational regulation.

The mRNA encoding the cAMP-responsive element (CRE)-binding protein (CREB) has been found to be translated within the axons of sympathetic neurons.[42] CREB is a transcription factor that has several functions. It has been studied in particular in neurons where is involved in the formation of long-term memories. The translation of axonally localized *CREB* mRNA occurs in response to nerve growth factor (NGF) stimulation. After local translation, the protein is trafficked to the cell body via retrograde transport. In neurons lacking *CREB* transcripts in their axons, both CRE-mediated transcription and neuronal survival induced by axonal application of NGF are abolished. These observations highlight a signaling function for the axonally synthesized CREB and the biological importance of the specific subcellular localization of *CREB* mRNA in axons. For instance, a signal-dependent synthesis and subsequent retrograde trafficking of transcription factors could enable specific transcriptional responses to signaling events occurring at distal axonal sites.

Local synthesis of specific proteins may also be necessary for the cotranslational incorporation of these proteins into target structures. For example, the assembly of nascent proteins into supramolecular complexes has been described for cytoskeletal proteins.[43] In a similar way, transcripts can be transported to and incorporated into specific cellular organelles. Several mRNAs coding for mitochondrial proteins are localized to mitochondrion-bound polysomes.[44] Similarly, some mRNAs were found to localize to the endoplasmic reticulum (ER). This organelle-targeting is thought to facilitate the cotranslational import of the encoded proteins into the respective organelles. The localization of mRNAs to the ER may further facilitate the sorting of integral membrane proteins to subdomains of the plasma membrane.[45] Both mitochondria and the ER are abundant at the post-synapse and have been shown to undergo activity-dependent changes in, for example, their localization.[46] As for localized mRNAs, LEs (cis-acting elements) have been identified in mitochondria-targeted mRNAs. In RNAs coding for proteins involved in mitochondrial biogenesis and/or activity, these elements could allow for transport to the vicinity of these organelles that are distributed along the neuronal processes reaching long distances. The LEs are then recognized by trans-acting factors, RBPs that subsequently recruit other factors including, for example, motor proteins. It is tempting to speculate that local protein synthesis followed by cotranslational targeting plays a role in modulating the functions of these organelles at the synapse. However, more experiments are needed to test this hypothesis.

RNA Transport: A Multistep Process

Where and when within a cell are mRNAs marked for transport? How does the process of RNA transport occur? In general, several phases can be distinguished[45,47] (Fig. 3). The nascent transcript must undergo a series of processes that take place within the nucleus. Most of the mRNAs that are transported into neuronal processes are packaged into large RNPs. It is accepted that RNP biogenesis starts with the beginning of transcription.[48] The nascent mRNA is bound by proteins that can be involved in many aspects of RNA biology: Splicing, quality control, stability and translational control. Some of these factors bind the mRNA in the nucleus and remain

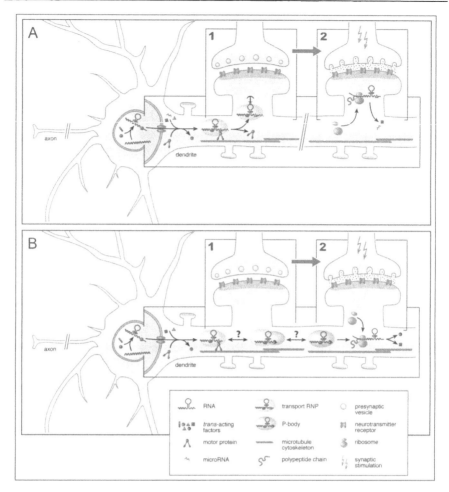

Figure 3. A) Model showing mRNA localization in mature mammalian neurons. Nascent mRNAs associate with a first set of proteins (trans-acting factors) which mark the transcripts for a specific subcellular localization. After nuclear export, these RNPs are believed to be remodeled by additional trans-acting factors assembling onto the mRNAs and others dissociating. This process leads to a transport-competent RNP (transport RNP) that moves into dendrites using molecular motors. For local protein synthesis to occur in a temporally and spatially restricted manner, the localized mRNAs must be kept translationally silent during transport. This can be achieved through several mechanisms, involving proteins and/or microRNAs that block translation. At their final destinations, e.g., the bases of dendritic spines, transport RNPs are believed to be anchored (inset 1). Stimulation leading to synaptic plasticity is thought to lift the translational block on the localized mRNAs and to lead to their translation (inset 2). The locally synthesized proteins can then be incorporated into the post-synaptic part of the synapse. B) Putative role of P-bodies in the control of local mRNA translation in mammalian neurons. The docking of P-bodies and transport RNPs, as observed in dendrites of mammalian neurons, might serve to transfer mRNAs from one particle-type to the other in a bidirectional fashion (inset 1). A relocation of an mRNA from a transport RNP into a P-body could, for instance, result in a more stable repression of the translation of this mRNA. Conversely, shifting an mRNA from a P-body back into a transport RNP might make the respective mRNA more amenable to the translation machinery. Synaptic stimulation, which was shown to cause the dispersion of at least some P-body markers, might liberate mRNAs from translational repression and allow the local synthesis of the corresponding proteins (inset 2). Figure reproduced from Zeitelhofer et al. RNA Biol 2008; 5:244-48,[132] with permission of Landes Bioscience.

associated with the transcript during transport. For example, eIF4AIII protein, a component of the exon-junction-complex, stays associated with neuronal mRNA granules and dendritically localized transcripts.[49] Similarly, the RBP Barentsz (now called Casc3), a key component in the nonsense mediated decay (NMD) machinery,[50] has been found to be associated with dendritically localized RNPs.[51,52] Finally, members belonging to the group of heterogeneous nuclear ribonucleoparticles (hnRNPs), which have a function in mRNA splicing, are critically involved also in cytoplasmic RNA trafficking.[53] The functional role(s) of these factors at the mechanistic level at synapses is still unclear. Whether they exert a direct or indirect role in RNA transport, e.g., mediating RNA stability and/or RNP assembly, needs further investigation.

For the initial association of the nascent mRNAs with proteins, the mRNAs are packaged into transport-competent particles. Part of this packaging is thought to occur already in the nucleus following or in parallel to transcription and nuclear processing of the RNAs.[54] Additional factors are then added in the cytoplasm subsequent to nuclear export.[55] The assembly of the RNPs relies both on the recognition of motifs in the RNA (cis-acting elements or zipcodes) by RBPs (trans-acting factors) and on protein-protein interactions between trans-acting factors. The identification of the zipcodes is often a difficult task.[56] Zipcodes span from a few to thousands of nucleotides in length and are recognized by the transport apparatus on the basis of both sequence and secondary or tertiary structure.

Whereas the mRNA 5′UTR contains elements important for translational control, the targeting sequences are usually located within the 3′UTR of the transported mRNAs.[32] This region also contains sequences involved in different processes of mRNA metabolism (e.g., stability, nuclear export, binding for microRNA and translational control). The low degree of conservation among the 3′UTR sequences renders the identification of these elements difficult when using bioinformatical approaches.[57] The nuclear priming of localized transcripts seems to be crucial for their proper subcellular localization.[58] However, experiments performed by microinjecting fluorescently labeled mRNA in neurons indicate that RNA localization can occur even upon injection into the cytoplasm.[59-61] After leaving the nucleus via active nuclear export, the RNPs can either form polysomes for translation or can undergo a different type of remodeling to form larger complexes known as RNA granules.

Several experimental approaches have been successfully used to identify the molecular components of transport RNA granules. A number of factors required for an active transport along the cytoskeleton (motor proteins, such as kinesis and dyneins), translational and stabilization controllers, e.g., the zipcode-binding protein1 (ZBP1), the fragile X mental retardation protein (FMRP), Pumilio2, the poly(A)-binding protein (PABP), splicing factors and proteins involved in RNA metabolism have been identified in neurons.[62] Which factors are necessary and sufficient to render RNA granules competent for transport is, however, still unknown.

Localized mRNAs that have been assembled into large RNPs are subsequently transported to their destinations, e.g., sites of local translation, using microtubules[63,64] (Figs. 3 and 4). Due to the neuronal morphology, the polarity of these components of the cytoskeleton is different compared to other cell types: of mixed polarity in proximal dendrites, unipolar in axons and distal dendrites.[29] In the latter case the plus ends point distally. Messenger RNAs assembled into RNPs are thought to be transported to neuronal processes by molecular motors.[27,64] Among these factors, microtubule-dependent motor proteins such as the kinesin superfamily proteins (KIFs) and the cytoplasmic dyneins seem to be essential.[65,66] Whereas dyneins mediate the retrograde transport processes (minus-end-directed movement), the kinesin superfamily proteins are usually involved in the anterograde movement (plus-end-directed motors). Kanai and colleagues showed that the conventional kinesin KIF5 is involved in the transport of a subset of mRNAs into dendrites.[67] Several components of the transport RNPs were isolated and identified upon immunoprecipitation of KIF5, including proteins involved in mRNA transport (e.g., Staufen, FMRP), translational control and protein helicases. Moreover, *CaMKIIα* and *Arc* mRNAs were isolated. Misexpression of KIF5 affects the direct transport of RNPs confirming the importance of kinesin proteins in the movement of RNA particles.

RNA granules are then anchored in the proximity of synapses.[68,69] Here, specific stimuli have to be integrated to induce the translation of the localized mRNAs.[26,70-72] How the mRNA is anchored when it reaches its target location in dendrites and which molecules are involved in this process is still unknown. Using an elegant experimental approach, Mili and colleagues recently showed that the APC-tumor suppressor protein is required for the accumulation of transcripts in the protrusions of migrating mouse fibroblasts.[73] The postulated function of APC is to mediate the interaction between mRNA and microtubules allowing for the anchoring of the transcripts to the cytoskeleton. Interestingly, a factor involved in translation elongation, EF1A, plays a similar role at the leading edge of migrating fibroblasts.[74] In this compartment, upon transport of the mRNA along the actin cytoskeleton, EF1A mediates a more stable interaction between the mRNA and the cytoskeleton. The EF1A protein, as well as its corresponding RNA, localizes in vivo to the dendrites of many neuronal cell types that exhibit LTP and LTD.[75] Moreover, its translation is locally regulated and increased upon pharmacological activation of mGluRs and by LTP. In particular, the peak in localization occurs during dendritogenesis, suggesting that local synthesis of EF1A plays an important role also in synaptogenesis. Whether EF1A is involved in the anchoring of the mRNA in dendrites of nerve cells is still unknown but very likely. Taken together, the mechanism leading to mRNA anchoring in neuronal processes and, in particular, at synapses is still poorly understood. However, all preliminary observations further link the process of transport with translational control.

Translational Control of Transported RNPs

For mRNA localization to be effective, the corresponding proteins must only be produced when the RNAs reach their final destinations. This requires translation to be shut off throughout the transport of the mRNAs. Several mechanisms have been proposed to account for this. The first clear demonstration of such a regulation has been provided by an elegant series of experiments a few years ago. The zip-code binding protein 1 (ZBP1) is an RBP that can determine the localization of β-actin mRNA at cell edges where cytoskeletal re-arrangement occurs both in fibroblasts and neurons, through binding to a conserved motif in the 3′UTR of the β-actin mRNA (Figs. 2 and 5). ZBP1 associates with this target mRNA in the nucleus and acts as a translational repressor once the mRNPs are exported in the cytoplasm until, when at destination, it is phosphorylated by the Src kinase in a tyrosine residue involved in ZBP1 mRNA binding.[36] This molecular event has been also recently shown to control growth cone turning responses to attractive or repulsive cues of developing axons, being triggered by the chemoattractant BDNF.[76] A similar mechanism of release of translational suppression by localized phosphorylation of RBPs has been reported for the ASH1 mRNA in yeast, a model to study spatiotemporal control of mRNA translation in vertebrates. Two RBPs binding the ASH1 mRNA, Khd1p and the Pumilio family member Puf6p, have been shown to be essential for its localized translation. ASH1 mRNA translation at the bud plasma membrane is allowed either by Khd1p release mediated by yeast CKI phosphorylation[77] or by Puf6p release mediated by yeast CK2 phosphorylation (see Chapter 5 by Niessing for details).[78]

But how do these RBPs inhibit translation before being released? Both ZBP1[36] and the yeast Pumilio protein[78] seem to act in preventing the formation of the 80S ribosome by blocking 60S subunit joining at the translation start site without affecting correct positioning of the 48S complex, as demonstrated by cell-free protein synthesis assays. Therefore they are supposed to avoid premature translation by stalling, in an unknown way, the very last phase of translation initiation, as was originally demonstrated for the hnRNP K and hnRNP E1 proteins with *lipoxygenase* mRNA in erythroid differentiation.[79] This model would imply, however, the presence of a stalled 48S complex in the translationally silenced transport mRNPs.

Other mechanisms could also be possible, even for the same transported mRNP. The use of a chemoattractant other than BDNF in *Xenopus* retinal growth cones, netrin-1, elicits ZBP1-dependent asymmetric translation and growth cone turning as well, but the way it activates translation is completely different, involving an asymmetric hyperphosphorylation and release of the eIF4E inhibitor 4E-BP mediated by the mTOR pathway.[37] This last mechanism is

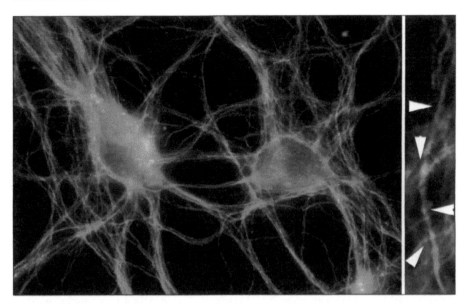

Figure 4. *MAP2* mRNA particles are transported along the cytoskeleton. MAP2 RNPs (in red) are found in the cell bodies and dendrites of 15 DIV cultured hippocampal neurons. The protein tubulin is shown in green. A high magnification image of *MAP2* mRNA in dendrites reveals a colocalization with microtubules stained with an anti-tubulin antibody (see arrowheads). (Image courtesy of Dr. Georgia Vendra.)

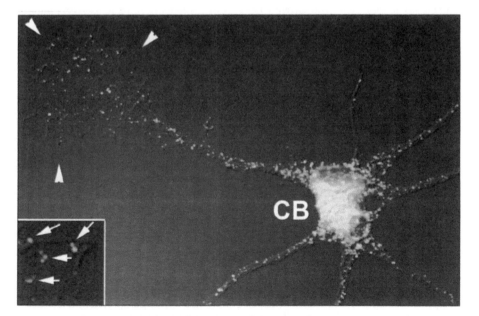

Figure 5. Subcellular localization of ZBP1 (Zipcode-binding protein 1) in neurons. Immunostaining for ZBP1 protein has been performed on a chick forebrain neuron at 4 DIV. ZBP particles (in green) are observed in the cell body (CB) as well as along the neurites. The arrowheads show an enlarged growth cone. Inset: Higher magnification showing ZBP1 particles within the growth cone. (Image courtesy of Gary Bassell and Kristy Welshhans.)

the best known way to activate cap-dependent translational at initiation, with the freeing of the rate-limiting cap binding protein eIF4E, the assembly of the eIF4F complex involving mRNA circularization (closed-loop) and its association with the 43S pre-initiation complex to form the 48S complex, which initiates 5'UTR scanning.[80] In this last case the pre-existing mRNA-specific translational repression is expected to be due to the presence of long, structured 5'UTRs in the mRNA, which compete poorly for efficient initiation due to difficult scanning, and are therefore much less translated than mRNAs with short and linear 5'UTRs.[81] Efficient eIF4F formation at these inefficiently scanned mRNAs would favour their translation.[82]

A more mechanistically explained way of preventing eIF4F-mediated mRNA circularization is invoked also to describe translational silencing due to the cytoplasmic polyadenylation element-binding protein 1 (CPEB1) RBP. CPEB1 was originally identified in *Xenopus* oocytes, and shown to promote mRNA polyadenylation and translation by binding to the conserved CPE motif.[83] Evidence of the role of CPEB1 in nervous system development came from studies of the growth cones of commissural axons in the chick spinal cord, where a CPE motif in the 3'UTR of the guidance receptor *EphA2* mRNA mediates its localized translation.[84] In neurite growth cones of developing hippocampal neurons, CPEB1 binds to *β-catenin* mRNA in a CPE-dependent manner, and activates its translation in response to the NT3 neurotrophin mediating process branching.[85] Moreover, CPEB1 localizes to synapses in mature mouse neurons, being an integral component of the PSD,[83] and is necessary for visual experience-dependent growth of *Xenopus* tectal dendrites.[86] In synapses of rat hippocampal neurons, CPEB1 is phosphorylated by Aurora A kinase in response of activation of the N-methyl-D-aspartate (NMDA) receptor,[87] while in the same cells during induction of synaptic plasticity it is also phosphorylated by CaMKII.[88]

How can these events be related to translational suppression and re-activation? From studies on the control of the dormant *cyclin B1* mRNA in *Xenopus* oocyte maturation, CPEB1 is part of a complex of proteins including symplekin, a scaffolding protein, the poly(A) polymerase GLD2, the PARN deadenylase[89] and Maskin, an eIF4E binding protein inducing a type of mRNA circularization which prevents eIF4F formation[90] and therefore represses translation. Following progesterone-induced CPEB1 phosphorylation PARN dissociates from the complex, allowing GLD2 to produce a net growth of the poly(A) tail,[91] which in turn induces dissociation of Maskin from eIF4E and activation of translation.[92] Progesterone-induced phosphorylation of Maskin contributes to this process.[93] This model has the historical merit of having proposed the formation of an inhibitory closed-loop between a 3'UTR binding RBP, a bridging protein, and eIF4E as a mechanism to repress translation in the mRNP. But the role of Maskin as a key bridging protein has been questioned in *Xenopus*, where at earlier stages of oocyte maturation CPEB1 is found in a complex where the inhibitory closed-loop could be formed by another protein, 4-ET, bound to eIF4E1b, an embryo-restricted paralogue of eIF4E. Remarkably, this arrangement is reproduced in at least three well-studied translational repressor complexes in *Drosophila* embryos, relative to the *Nanos, Oskar, Caudal* and *Hunckback* mRNAs. In all these cases the mRNA cap-binding protein is an eIF4E paralogue, 4EHP, or eIF4E itself (there are 8 eIF4E family members in *Drosophila*), the 3'UTR binding protein is, respectively, Smaug, Bruno, Bicoid and Pumilio, while the bridging protein is the 4-ET ortholog Cup for *Nanos* and *Oskar*, Bicoid itself for *Caudal* and Pumilio itself associated to Nanos and Brat for *Hunckback*.[80] Is there a preferred cellular site for this translational repression by production of inhibitory closed loops to take place, and to what extent this model could be extended from the early phases of invertebrate and vertebrate ontogenesis to the vertebrate nervous system? We can attempt an answer to both these questions by starting with the demonstration, again in *Xenopus* oocytes, of additional corepressor proteins present in the CEBP1/4-ET/eIF4Eb1 complex. These proteins are the RNA helicase Xp54/RCK and the RBPs PAT1 and RAP55.[94] All of them are known, structural components of the so-called P-bodies (mRNA processing bodies), discrete cytoplasmic domains ubiquitously present in eukaryotic cells where mRNA degradation, silencing, quality control and possibly translational repression are colocalized and integrated.[95,96] In neurons, P-bodies appear to be interesting candidates for controlling mRNA translation and turnover of localized mRNAs (Fig. 3B). In *Drosophila* sensory neurons the P-body markers Xp54/RCK, DCP1, UPF1 and AGO2 were detected either in the somata or in dendrites,[97] and P-bodies markers are also clearly present in dendrites of cultured rat hippocampal neurons, responding dynamically to synaptic stimulation with disassembly or loss of marker components.[98-100] While in *Drosophila* P-bodies colocalize with a marker of transport mRNPs, Staufen,[97] in rodent neurons they are not constantly associated or are cotransported but dynamically interact with transport mRNPs by docking.[100]

Again in *Drosophila*[97] and in human neuroblastoma cells,[101] P-bodies colocalize with granules containing FMRP, another well-studied protein involved in translational repression in neurons.[102] Mutations in the gene encoding FMRP are the cause of the most prevalent-type of mental retardation, and of a form of autism (see Chapter 3 by Rogelj et al for details on how mutations in FMRP can influence its function). In neurons FMRP is localized in axons and dendrites,[103] and is also a component of mRNPs transported to distant axonal and to synaptic locations along microtubules, with an active role in the kinesin-dependent transport process following glutamatergic stimulation.[104] But FMRP is also localized in polysomes,[105] and can repress translation in vitro,[106,107] while at synapses of FMRP gene knockout mice an excess of MAP1b, CaMKIIα and Arc proteins and polysomally loaded mRNAs is present.[108] Glutamatergic stimulation in the same mice does not induce the normal recruitment of *CaMKIIα, GluR1/2* and *PSD-95* mRNAs to polysomes,[109] showing that FMRP is both a translational repressor and a metabotropic glutamate receptor type-I (mGluR) induced translational activator. Recent studies are beginning to elucidate the molecular mechanisms of this complex activity. mGluR stimulation induces early dephosphorylation of FMRP by the PP2A phosphatase and late rephosphorylation mediated by the S6K1 kinase, requiring signaling inputs through the mTOR pathway and promoting at least translational repression of SAPAP3, a synapse-associated FMRP target mRNA.[110,111] It is, however, still completely unclear how this phosphorylation event acts in promoting translational silencing, and where it happens with respect to the multiple compartmentalizations (transport mRNP granules, P-bodies, polysomes) of FMRP. A recent interesting study proposes an inhibitory closed-loop like mechanism for FMRP-mediated translational repression. CYFIP1 is a FMRP interacting protein[112] whose *Drosophila* ortholog mutation affects, like FMRP mutations, axon and synapse morphology.[113] CYFIP1 binds also to eIF4E through an interacting domain related to those described for other eIF4E binding proteins, including 4E-BP1 and eIF4G and, in response to mGluR of BDNF stimulation, is dissociated from eIF4E at the synapse possibly releasing translational repression.[114]

Finally, the complex role of FMRP in translation of localized mRNAs calls on the scene another possibly general component of the translational repressor machinery, the microRNA (miRNA) class of small noncoding RNA. Early identification of microRNAs in the FMRP/mRNA recognition complex[115] has been later confirmed, with the inclusion of components of the microRNA processing and recognition complex, such as DICER, AGO1 and AGO2.[116] In mammals pre-microRNA are trascribed and processed in the nucleus, then released into the cytoplasm where they are cut into short, double stranded duplexes by the DICER complex, which are then separated and associated to AGO2 and the target mRNAs in their 3'UTR. Partial complementarity of the miRNA to its target mRNAs generally leads to a translational repression, while full matching triggers AGO2 endonuclease activity, leading to cleavage and complete degradation of the target mRNA.[117] Recent work clearly demonstrated that FMRP associates with a number of miRNA species and functionally co-operates with miR-125b and AGO1 in repressing NMDA receptor subunit NR2A mRNA translation.[118] The novelty of the study relies on the demonstration that a translational regulator RBP acts together with the miRNA pathway on specific target mRNAs during synapse dynamics. Besides this interaction, miRNAs are now recognized as controllers of the state of neuronally localized mRNAs with functional consequences, several of them having been involved for example in neuronal development. We direct the reader to comprehensive recent reviews about this topic.[119-121]

Another mechanism of translational inhibition is provided by the RBP Musashi 1. In the mammalian nervous system Musashi 1 is present in neural stem cells, where it contributes to the maintenance of stemness by repressing translation of an inhibitor of Notch signaling, Numb.[122] Musashi 1 binds to PABP competing with PABP binding to eIF4G, and therefore probably not allowing formation of the closed-loop.[123] Apparently, this does not affect assembly of the 48S complex but instead inhibits recruitment of the 60S ribosomal subunit and translation start.[123]

A final way by which translation is activated in neuronal mRNPs is provided by the ELAV RBPs. This is a family of four highly conserved RBPs, three of which are specifically expressed along the neuronal differentiation lineage and are considered master regulators of this cell program.[124] ELAV proteins positively regulate gene expression of the target mRNAs, possibly acting as mRNA stabilization factors,[125,126] even if in several cases they have been shown to behave as translational enhancers,[127-129] and are well known for being polysome-associated proteins.[130] A recent study suggests a mechanism of translational activation for the ELAV protein HuD, which involves both poly(A)-binding and interaction with eIF4A, the RNA helicase, which is part of the eIF4F complex and responsible for efficient 5'UTR mRNA scanning.[131] To our knowledge, this is the first example of sequence-dependent translational activation exerted on the eIF4F complex in an alternative way to eIF4E release.

In summary, two main molecular mechanisms depending on RBPs, and therefore exquisitely sequence-specific, are emerging in the activation of translation for localized mRNAs: The first, highly conserved in evolution and originally described during development, involves formation of a translationally unproductive closed-loop, sequestering the cap binding protein eIF4E and competing with the eIF4F-induced mRNA circularization which brings to the 48S complex formation. The second, instead, acts on this complex contrasting in an unknown way the last preparatory process to translation initiation, joining of the 60S ribosomal subunit. Phosphorylation is at the moment the only protein modification proved to promote translation relieving for both these types of block.

Conclusion

Defects in mRNA localization and translational regulation can have severe consequences for cellular development and function. Such defects underlie, for instance, several human diseases, including the fragile X mental retardation syndrome (the most common form of heritable mental retardation), spinal muscular and myotonic atrophies, paraneoplastic opsoclonus-myoclonus and others. As our understanding of the molecular mechanisms underlying mRNA localisation and translational control becomes better, more human pathologies will likely be found to be caused by errors in these processes.

Since the 1960s, numerous labs have contributed to dissecting the molecular mechanisms involved in subcellular mRNA localization in neurons as well as other cell types. However, key questions are still open: Do RNPs that are involved in the transport of different mRNAs share common molecular components? What is the complete molecular composition of transport RNPs and which are the functions of the individual components? How is translational control achieved during the transport? One key feature of mRNA localization is that localization precedes translation. Little is still known about the mechanisms by which the translation of silenced mRNAs is activated once localized transcripts have reached their final destinations, e.g. at synapses that have received signals inducing plasticity? And how is the degradation of transported RNAs regulated locally?

The identification in the last years of cytoplasmic mRNA storage compartments, which are a universal feature of eukaryotic cells, and the growing appreciation of a widespread uncoupling between transcriptional and translational dynamics suggest that we are only beginning to explore the ways by which neurons realize their functional diversity by selectively controlling translation. In order to proceed along this path new methods are needed which will allow a better definition of the molecular roles of the hundreds of uncharacterized RBPs and of the thousands of non-coding RNAs which could be key players of this process.

Acknowledgments

We are grateful to Drs. Gary Bassell (Emory University School of Medicine) and Georgia Vendra (Medical University of Vienna) for kindly providing figures. The authors wish to thank Dr. Sheref Mansy for critical reading of the manuscript. We apologize to all authors whose work could not be cited due to space restrictions. This work was supported by the Italian Neuroblastoma Foundation (to A.Q.) and by a CARITRO (Cassa di Risparmio di Trento e Rovereto) Foundation grant (to P.M.). The financial support of the University of Trento (Progetto Biotecnologie) is gratefully acknowledged (to A.Q. and P.M.).

References

1. Bodian D. An Electron-microscopic study of the monkey spinal cord. I. Fine structure of normal motor column. Ii. Effects of retrograde chromatolysis. Iii. Cytologic effects of mild and virulent poliovirus infection. Bull Johns Hopkins Hosp 1964; 114:13-119.
2. Steward O, Levy WB. Preferential localization of polyribosomes under the base of dendritic spines in granule cells of the dentate gyrus. J Neurosci 1982; 2:284-291.
3. Burgin KE, Waxham MN, Rickling S et al. In situ hybridization histochemistry of Ca2+/calmodulin-dependent protein kinase in developing rat brain. J Neurosci 1990; 10:1788-1798.
4. Garner CC, Tucker RP, Matus A. Selective localization of messenger RNA for cytoskeletal protein MAP2 in dendrites. Nature 1988; 336:674-677.
5. Böckers TM, Segger-Junius M, Iglauer P et al. Differential expression and dendritic transcript localization of Shank family members: identification of a dendritic targeting element in the 3' untranslated region of Shank1 mRNA. Mol Cell Neurosci 2004; 26:182-190.
6. Kindler S, Muller R, Chung WJ et al. Molecular characterization of dendritically localized transcripts encoding MAP2. Brain Res Mol Brain Res 1996; 36:63-69.
7. Blichenberg A, Schwanke B, Rehbein M et al. Identification of a cis-acting dendritic targeting element in MAP2 mRNAs. J Neurosci 1999; 19:8818-8829.
8. Blichenberg A, Rehbein M, Müller R et al. Identification of a cis-acting dendritic targeting element in the mRNA encoding the alpha subunit of Ca2+/calmodulin-dependent protein kinase II. Eur J Neurosci 2001; 13:1881-1888.
9. Mori Y, Imaizumi K, Katayama T et al. Two cis-acting elements in the 3' untranslated region of alpha-CaMKII regulate its dendritic targeting. Nat Neurosci 2000; 3:1079-1084.
10. Miller S, Yasuda M, Coats JK et al. Disruption of dendritic translation of CaMKIIalpha impairs stabilization of synaptic plasticity and memory consolidation. Neuron 2002; 36:507-519.
11. Steward O, Wallace CS, Lyford GL et al. Synaptic activation causes the mRNA for the IEG Arc to localize selectively near activated postsynaptic sites on dendrites. Neuron 1998; 21:741-751.
12. Steward O, Worley PF. A cellular mechanism for targeting newly synthesized mRNAs to synaptic sites on dendrites. Proc Natl Acad Sci USA 2001; 98:7062-7068.
13. Tongiorgi E, Righi M, Cattaneo A. Activity-dependent dendritic targeting of BDNF and TrkB mRNAs in hippocampal neurons. J Neurosci 1997; 17:9492-9505.
14. Huang F, Chotiner JK, Steward O. The mRNA for elongation factor 1alpha is localized in dendrites and translated in response to treatments that induce long-term depression. J Neurosci 2005; 25:7199-7209.
15. Zhong J, Zhang T, Bloch LM. Dendritic mRNAs encode diversified functionalities in hippocampal pyramidal neurons. BMC Neurosci 2006; 7:17.
16. Bolognani F, Merhege MA, Twiss J et al. Dendritic localization of the RNA-binding protein HuD in hippocampal neurons: association with polysomes and upregulation during contextual learning. Neurosci Lett 2004; 371:152-157.
17. Muslimov IA, Nimmrich V, Hernandez AI et al. Dendritic transport and localization of protein kinase Mzeta mRNA: implications for molecular memory consolidation. J Biol Chem 2004; 279:52613-52622.
18. Grange J, Boyer V, Fabian-Fine R et al. Somatodendritic localization and mRNA association of the splicing regulatory protein Sam68 in the hippocampus and cortex. J Neurosci Res 2004; 75:654-666.
19. Falley K, Schütt J, Iglauer P et al. Shank1 mRNA: dendritic transport by kinesin and translational control by the 5'untranslated region. Traffic 2009; 10:844-857.
20. Lim S, Naisbitt S, Yoon J et al. Characterization of the Shank family of synaptic proteins. Multiple genes, alternative splicing, and differential expression in brain and development. J Biol Chem 1999; 274:29510-29518.
21. Mohr E, Prakash N, Vieluf K et al. Vasopressin mRNA localization in nerve cells: characterization of cis-acting elements and trans-acting factors. Proc Natl Acad Sci USA 2001; 98:7072-7079.
22. Mohr E, Richter D. Subcellular vasopressin mRNA trafficking and local translation in dendrites. J Neuroendocrinol 2004; 16:333-339.
23. Sakagami H, Matsuya S, Nishimura H et al. Somatodendritic localization of the mRNA for EFA6A, a guanine nucleotide exchange protein for ARF6, in rat hippocampus and its involvement in dendritic formation. Eur J Neurosci 2004; 19:863-870.
24. Hernandez-Deviez DJ, Casanova JE, Wilson JM. Regulation of dendritic development by the ARF exchange factor ARNO. Nat Neurosci 2002; 5:623-624.
25. Holt CE, Bullock SL. Subcellular mRNA localization in animal cells and why it matters. Science 2009; 326:1212-1216.
26. Wang DO, Martin KC, Zukin RS. Spatially restricting gene expression by local translation at synapses. Trends Neurosci 2010; 33:173-182.
27. Bramham CR, Wells DG. Dendritic mRNA: transport, translation and function. Nat Rev Neurosci 2007; 8:776-789.

28. Sossin WS, DesGroseillers L. Intracellular trafficking of RNA in neurons. Traffic 2006; 7:1581-1589.
29. Hirokawa N. mRNA transport in dendrites: RNA granules, motors, and tracks. J Neurosci 2006; 26:7139-7142.
30. Mikl M, Vendra G, Doyle M et al. RNA localization in neurite morphogenesis and synaptic regulation: current evidence and novel approaches. J Comp Physiol A Neuroethol Sens Neural Behav Physiol 2010; 196:321-334.
31. Meignin C, Davis I. Transmitting the message: intracellular mRNA localization. Curr Opin Cell Biol 2010; 22:112-119.
32. Martin KC, Ephrussi A. mRNA localization: gene expression in the spatial dimension. Cell 2009; 136:719-730.
33. Sotelo-Silveira JR, Calliari A, Kun A et al. RNA trafficking in axons. Traffic 2006; 7:508-515.
34. Yoo S, van Niekerk EA, Merianda TT et al. Dynamics of axonal mRNA transport and implications for peripheral nerve regeneration. Exp Neurol 2010; 223:19-27.
35. Bassell GJ, Zhang H, Byrd AL et al. Sorting of beta-actin mRNA and protein to neurites and growth cones in culture. J Neurosci 1998; 18:251-265.
36. Hüttelmaier S, Zenklusen D, Lederer M et al. Spatial regulation of beta-actin translation by Src-dependent phosphorylation of ZBP1. Nature 2005; 438:512-515.
37. Leung KM, van Horck FP, Lin AC et al. Asymmetrical beta-actin mRNA translation in growth cones mediates attractive turning to netrin-1. Nat Neurosci 2006; 9:1247-1256.
38. Yao J, Sasaki Y, Wen Z et al. An essential role for beta-actin mRNA localization and translation in Ca2+-dependent growth cone guidance. Nat Neurosci 2006; 9:1265-1273.
39. Zhang HL, Eom T, Oleynikov Y et al. Neurotrophin-induced transport of a beta-actin mRNP complex increases beta-actin levels and stimulates growth cone motility. Neuron 2001; 31:261-275.
40. Aronov S, Aranda G, Behar L et al. Visualization of translated tau protein in the axons of neuronal P19 cells and characterization of tau RNP granules. J Cell Sci 2002; 115:3817-3827.
41. Bi J, Tsai NP, Lin YP et al. Axonal mRNA transport and localized translational regulation of kappa-opioid receptor in primary neurons of dorsal root ganglia. Proc Natl Acad Sci USA 2006; 103:19919-19924.
42. Andreassi C, Zimmermann C, Mitter R et al. An NGF-responsive element targets myo-inositol mono-phosphatase-1 mRNA to sympathetic neuron axons. Nat Neurosci 2010; 13:291-301.
43. Chang L, Shav-Tal Y, Trcek T et al. Assembling an intermediate filament network by dynamic cotranslation. J Cell Biol 2006; 172:747-758.
44. Corral-Debrinski M. mRNA specific subcellular localization represents a crucial step for fine-tuning of gene expression in mammalian cells. Biochim Biophys Acta 2007; 1773:473-475.
45. Jansen RP. mRNA localization: message on the move. Nat Rev Mol Cell Biol 2001; 2:247-256.
46. Li Z, Okamoto K, Hayashi Y et al. The importance of dendritic mitochondria in the morphogenesis and plasticity of spines and synapses. Cell 2004; 119:873-887.
47. Wilhelm JE, Vale RD. RNA on the move: the mRNA localization pathway. J Cell Biol 1993; 123:269-274.
48. Luna R, Gaillard H, Gonzalez-Aguilera C et al. Biogenesis of mRNPs: integrating different processes in the eukaryotic nucleus. Chromosoma 2008; 117:319-331.
49. Giorgi C, Yeo GW, Stone ME et al. The EJC factor eIF4AIII modulates synaptic strength and neuronal protein expression. Cell 2007; 130:179-191.
50. Palacios IM, Gatfield D, St Johnston D et al. An eIF4AIII-containing complex required for mRNA localization and nonsense-mediated mRNA decay. Nature 2004; 427:753-757.
51. Macchi P, Kroening S, Palacios IM et al. Barentsz, a new component of the Staufen-containing ribonucleoprotein particles in mammalian cells, interacts with Staufen in an RNA-dependent manner. J Neurosci 2003; 23:5778-5788.
52. Van Eeden FJ, Palacios IM, Petronczki M et al. Barentsz is essential for the posterior localization of oskar mRNA and colocalizes with it to the posterior pole. J Cell Biol 2001; 154:511-523.
53. Percipalle P, Raju CS, Fukuda N. Actin-associated hnRNP proteins as transacting factors in the control of mRNA transport and localization. RNA Biol 2009; 6:171-174.
54. Kiebler MA, Jansen RP, Dahm R et al. A putative nuclear function for mammalian Staufen. Trends Biochem Sci 2005; 30:228-231.
55. Lopez de Heredia M, Jansen RP. mRNA localization and the cytoskeleton. Curr Opin Cell Biol 2004; 16:80-85.
56. Jambhekar A, Derisi JL. Cis-acting determinants of asymmetric, cytoplasmic RNA transport. RNA 2007; 13:625-642.
57. Andreassi C, Riccio A. To localize or not to localize: mRNA fate is in 3'UTR ends. Trends Cell Biol 2009; 19:465-474.
58. Trcek T, Singer RH. The cytoplasmic fate of an mRNP is determined cotranscriptionally: exception or rule? Genes Dev 2010; 24:1827-1831.

59. Tübing F, Vendra G, Mikl M et al. Dendritically localized transcripts are sorted into distinct ribonu-cleoprotein particles that display fast directional motility along dendrites of hippocampal neurons. J Neurosci 2010; 30:4160-4170.

60. ShanJ, Munro TP, Barbarese E et al. A molecular mechanism for mRNA trafficking in neuronal dendrites. J Neurosci 2003; 23:8859-8866.

61. Gao Y, Tatavarty V, Korza G et al. Multiplexed dendritic targeting of alpha calcium calmodulin-dependent protein kinase II, neurogranin, and activity-regulated cytoskeleton-associated protein RNAs by the A2 pathway. Mol Biol Cell 2008; 19:2311-2327.

62. Elvira G, Wasiak S, Blandford V et al. Characterization of an RNA granule from developing brain. Mol Cell Proteomics 2006; 5:635-651.

63. Hirokawa N, Takemura R. Molecular motors and mechanisms of directional transport in neurons. Nat Rev Neurosci 2005; 6:201-214.

64. Kiebler MA, Bassell GJ. Neuronal RNA granules: movers and makers. Neuron 2006; 51:685-690.

65. Carson JH, Cui H, Barbarese E. The balance of power in RNA trafficking. Curr Opin Neurobiol 2001; 11:558-563.

66. Carson JH, Cui H, Krueger W et al. RNA trafficking in oligodendrocytes. Results Probl Cell Differ 2001; 34:69-81.

67. Kanai Y, Dohmae N, Hirokawa N. Kinesin transports RNA: isolation and characterization of an RNA-transporting granule. Neuron 2004; 43:513-525.

68. Dahm R, Macchi P. Human pathologies associated with defective RNA transport and localization in the nervous system. Biol Cell 2007; 99:649-661.

69. Dahm R, Kiebler M, Macchi P. RNA localisation in the nervous system. Semin Cell Dev Biol 2007; 18:216-223.

70. Sutton MA, Schuman EM. Dendritic protein synthesis, synaptic plasticity, and memory. Cell 2006; 127:49-58.

71. Schuman EM, Dynes JL, Steward O. Synaptic regulation of translation of dendritic mRNAs. J Neurosci 2006; 26:7143-7146.

72. Steward O, Schuman EM. Protein synthesis at synaptic sites on dendrites. Annu Rev Neurosci 2001; 24:299-325.

73. Mili S, Moissoglu K, Macara IG. Genome-wide screen reveals APC-associated RNAs enriched in cell protrusions. Nature 2008; 453:115-119.

74. Liu G, Grant WM, Persky D et al. Interactions of elongation factor 1alpha with F-actin and beta-actin mRNA: implications for anchoring mRNA in cell protrusions. Mol Biol Cell 2002; 13:579-592.

75. Tsokas P, Grace EA, Chan P et al. Local protein synthesis mediates a rapid increase in dendritic elonga-tion factor 1A after induction of late long-term potentiation. J Neurosci 2005; 25:5833-5843.

76. Sasaki Y, Welshhans K, Wen Z et al. Phosphorylation of zipcode binding protein 1 is required for brain-derived neurotrophic factor signaling of local beta-actin synthesis and growth cone turning. J Neurosci 2010; 30:9349-9358.

77. Paquin N, Ménade M, Poirier G et al. Local activation of yeast ASH1 mRNA translation through phosphorylation of Khd1p by the casein kinase Yck1p. Mol Cell 2007; 26:795-809.

78. Deng Y, Singer RH, Gu W. Translation of ASH1 mRNA is repressed by Puf6p-Fun12p/eIF5B interac-tion and released by CK2 phosphorylation. Genes Dev 2008; 22:1037-1050.

79. Ostareck DH, Ostareck-Lederer A, Shatsky IN et al. Lipoxygenase mRNA silencing in erythroid differentiation: The 3'UTR regulatory complex controls 60S ribosomal subunit joining. Cell 2001; 104:281-290.

80. Jackson RJ, Hellen CU, Pestova TV. The mechanism of eukaryotic translation initiation and principles of its regulation. Nat Rev Mol Cell Biol 2010; 11:113-127.

81. Koromilas AE, Lazaris-Karatzas A, Sonenberg N. mRNAs containing extensive secondary structure in their 5' non-coding region translate efficiently in cells overexpressing initiation factor eIF-4E. EMBO J 1992; 11:4153-4158.

82. Gingras AC, Raught B, Sonenberg N. mTOR signaling to translation. Curr Top Microbiol Immunol 2004; 279:169-197.

83. Wu L, Wells D, Tay J et al. CPEB-mediated cytoplasmic polyadenylation and the regulation of experi-ence-dependent translation of alpha-CaMKII mRNA at synapses. Neuron 1998; 21:1129-1139.

84. Brittis PA, Lu Q, Flanagan JG. Axonal protein synthesis provides a mechanism for localized regulation at an intermediate target. Cell 2002; 110:223-235.

85. Kundel M, Jones KJ, Shin CY et al. Cytoplasmic polyadenylation element-binding protein regulates neurotrophin-3-dependent beta-catenin mRNA translation in developing hippocampal neurons. J Neurosci 2009; 29:13630-13639.

86. Bestman JE, Cline HT. The RNA binding protein CPEB regulates dendrite morphogenesis and neuronal circuit assembly in vivo. Proc Natl Acad Sci USA 2008; 105:20494-20499.

87. Huang YS, Jung MY, Sarkissian M et al. J.D. N-methyl-D-aspartate receptor signaling results in Aurora kinase-catalyzed CPEB phosphorylation and alpha CaMKII mRNA polyadenylation at synapses. EMBO J 2002; 21:2139-2148.
88. Atkins CM, Nozaki N, Shigeri Y et al. Cytoplasmic polyadenylation element binding protein-dependent protein synthesis is regulated by calcium/calmodulin-dependent protein kinase II. J Neurosci 2004; 24:5193-5201.
89. Barnard DC, Ryan K, Manley JL et al. Symplekin and xGLD-2 are required for CPEB-mediated cytoplasmic polyadenylation. Cell 2004; 119:641-651.
90. Stebbins-Boaz B, Cao Q, de Moor CH et al. Maskin is a CPEB-associated factor that transiently interacts with eIF-4E. Mol Cell 1999; 4:1017-1027.
91. Kim JH, Richter JD. Opposing polymerase-deadenylase activities regulate cytoplasmic polyadenylation. Mol Cell 2006; 24:173-183.
92. Cao Q, Richter JD. Dissolution of the maskin-eIF4E complex by cytoplasmic polyadenylation and poly(A)-binding protein controls cyclin B1 mRNA translation and oocyte maturation. EMBO J 2002; 21:3852-3862.
93. Barnard DC, Cao Q, Richter JD. Differential phosphorylation controls Maskin association with eukaryotic translation initiation factor 4E and localization on the mitotic apparatus. Mol Cell Biol 2005; 25:7605-7615.
94. Minshall N, Reiter MH, Weil D et al. CPEB interacts with an ovary-specific eIF4E and 4E-T in early Xenopus oocytes. J Biol Chem 2007; 282:37389-37401.
95. Balagopal V, Parker R. Polysomes, P bodies and stress granules: states and fates of eukaryotic mRNAs. Curr Opin Cell Biol 2009; 21:403-408.
96. Eulalio A, Behm-Ansmant I, Izaurralde E. P-bodies: at the crossroads of posttranscriptional pathways. Nat Rev Mol Cell Biol 2007; 8:9-22.
97. Barbee SA, Estes PS, Cziko AM et al. Staufen- and FMRP-containing neuronal RNPs are structurally and functionally related to somatic P bodies. Neuron 2006; 52:997-1009.
98. Cougot N, Bhattacharyya SN, Tapia-Arancibia L et al. Dendrites of mammalian neurons contain specialized P-body-like structures that respond to neuronal activation. J Neurosci 2008; 28:13793-13804.
99. Vessey JP, Vaccani A, Xie Y et al. Dendritic localization of the translational repressor Pumilio 2 and its contribution to dendritic stress granules. J Neurosci 2006; 26:6496-6508.
100. Zeitelhofer M, Karra D, Macchi P et al. Dynamic interaction between P-bodies and transport ribonucleoprotein particles in dendrites of mature hippocampal neurons. J Neurosci 2008; 28:7555-7562.
101. Lee EK, Kim HH, Kuwano Y et al. hnRNP C promotes APP translation by competing with FMRP for APP mRNA recruitment to P bodies. Nat Struct Mol Biol 2010; 17:732-739.
102. Bassell GJ, Warren ST. Fragile X syndrome: loss of local mRNA regulation alters synaptic development and function. Neuron 2008; 60:201-214.
103. Antar LN, Li C, Zhang H et al. Local functions for FMRP in axon growth cone motility and activity-dependent regulation of filopodia and spine synapses. Mol Cell Neurosci 2006; 32:37-48.
104. Dictenberg JB, Swanger SA, Antar LN et al. A direct role for FMRP in activity-dependent dendritic mRNA transport links filopodial-spine morphogenesis to fragile X syndrome. Dev Cell 2008; 14:926-939.
105. Feng Y, Absher D, Eberhart DE et al. FMRP associates with polyribosomes as an mRNP, and the I304N mutation of severe fragile X syndrome abolishes this association. Mol Cell 1997; 1:109-118.
106. Laggerbauer B, Ostareck D, Keidel EM et al. Evidence that fragile X mental retardation protein is a negative regulator of translation. Hum Mol Genet 2001; 10:329-338.
107. Li Z, Zhang Y, Ku L et al. The fragile X mental retardation protein inhibits translation via interacting with mRNA. Nucleic Acids Res 2001; 29:2276-2283.
108. Zalfa F, Giorgi M, Primerano B et al. The fragile X syndrome protein FMRP associates with BC1 RNA and regulates the translation of specific mRNAs at synapses. Cell 2003; 112:317-327.
109. Muddashetty RS, Kelic S, Gross C et al. Dysregulated metabotropic glutamate receptor-dependent translation of AMPA receptor and postsynaptic density-95 mRNAs at synapses in a mouse model of fragile X syndrome. J Neurosci 2007; 27:5338-5348.
110. Narayanan U, Nalavadi V, Nakamoto M et al. FMRP phosphorylation reveals an immediate-early signaling pathway triggered by group I mGluR and mediated by PP2A. J Neurosci 2007; 27:14349-14357.
111. Narayanan U, Nalavadi V, Nakamoto M et al. S6K1 phosphorylates and regulates fragile X mental retardation protein (FMRP) with the neuronal protein synthesis-dependent mammalian target of rapamycin (mTOR) signaling cascade. J Biol Chem 2008; 283:18478-18482.
112. Schenck A, Bardoni B, Moro A et al. A highly conserved protein family interacting with the fragile X mental retardation protein (FMRP) and displaying selective interactions with FMRP-related proteins FXR1P and FXR2P. Proc Natl Acad Sci USA 2001; 98:8844-8849.
113. Schenck A, Bardoni B, Langmann C et al. CYFIP/Sra-1 controls neuronal connectivity in Drosophila and links the Rac1 GTPase pathway to the fragile X protein. Neuron 2003; 38:887-898.

114. Napoli I, Mercaldo V, Boyl PP et al. The fragile X syndrome protein represses activity-dependent translation through CYFIP1, a new 4E-BP. Cell 2008; 134:1042-1054.
115. Jin P, Zarnescu DC, Ceman S et al. Biochemical and genetic interaction between the fragile X mental retardation protein and the microRNA pathway. Nat Neurosci 2004; 7:113-117.
116. Cheever A, Ceman S. Translation regulation of mRNAs by the fragile X family of proteins through the microRNA pathway. RNA Biol 2009; 6:175-178.
117. Eulalio A, Huntzinger E, Izaurralde E. Getting to the root of miRNA-mediated gene silencing. Cell 2008; 132:9-14.
118. Edbauer D, Neilson JR, Foster KA et al. Regulation of synaptic structure and function by FMRP-associated microRNAs miR-125b and miR-132. Neuron 2010; 65:373-384.
119. Fineberg SK, Kosik KS, Davidson BL. MicroRNAs potentiate neural development. Neuron 2009; 64:303-309.
120. Saba R, Schratt GM. MicroRNAs in neuronal development, function and dysfunction. Brain Res 2010; 1338:3-13.
121. Schratt G. microRNAs at the synapse. Nat Rev Neurosci 2009; 10:842-849.
122. Imai T, Tokunaga A, Yoshida T et al. The neural RNA-binding protein Musashi1 translationally regulates mammalian numb gene expression by interacting with its mRNA. Mol Cell Biol 2001; 21:3888-3900.
123. Kawahara H, Imai T, Imataka H et al. Neural RNA-binding protein Musashi1 inhibits translation initiation by competing with eIF4G for PABP. J Cell Biol 2008; 181:639-653.
124. Pascale A, Amadio M, Quattrone A. Defining a neuron: neuronal ELAV proteins. Cell Mol Life Sci 2008; 65:128-140.
125. Brennan CM, Steitz JA. HuR and mRNA stability. Cell Mol Life Sci 2001; 58:266-277.
126. Keene JD. Ribonucleoprotein infrastructure regulating the flow of genetic information between the genome and the proteome. Proc Natl Acad Sci USA 2001; 98:7018-7024.
127. Antic D, Lu N, Keen JD. ELAV tumor antigen, Hel-N1, increases translation of neurofilament M mRNA and induces formation of neurites in human teratocarcinoma cells. Genes Dev 1999; 13:449-461.
128. Jain RG, Andrews LG, McGowan KM et al. Ectopic expression of Hel-N1, an RNA-binding protein, increases glucose transporter (GLUT1) expression in 3T3-L1 adipocytes. Mol Cell Biol 1997; 17:954-962.
129. Mazan-Mamczarz K, Galbán S, López de Silanes I et al. RNA-binding protein HuR enhances p53 translation in response to ultraviolet light irradiation. Proc Natl Acad Sci USA 2003; 100:8354-8359.
130. Antic D, Keene JD. Messenger ribonucleoprotein complexes containing human ELAV proteins: interactions with cytoskeleton and translational apparatus. J Cell Sci 1998; 111:183-197.
131. Fukao A, Sasano Y, Imataka H et al. The ELAV protein HuD stimulates cap-dependent translation in a Poly(A)- and eIF4A-dependent manner. Mol Cell 2009; 36:1007-1017.
132. Zeitelhofer M, Macchi P, Dahm R. Perplexing bodies: The putative roles of P-bodies in neurons. RNA Biol 2008; 5:244-248.

CHAPTER 5

RNA-Binding Proteins in Fungi and Their Role in mRNA Localization

Dierk Niessing*

Abstract

MRNA localization is a widely used mechanism for the spatial and temporal control of gene expression. In higher eukaryotes, the mRNA-transport machinery has a complexity that renders a mechanistic understanding at the molecular level difficult. During the last 15 years, studies in fungi have proven to be attractive alternatives. They are experimentally more accessible and the organization of their mRNA-transport particles is less complex. Here, our current understanding of mRNA transport in fungi will be summarized. The highly specific mRNA localization observed in the budding yeast *Saccharomyces cerevisiae* will be compared with mRNA localization in the human pathogen *Candida albicans* and the less specific transport of transcripts in the plant pathogen *Ustilago maydis*.

Introduction

One of the great advantages of yeast is its simplicity. Although being a eukaryotic organism, its cellular machinery often relies on fewer factors and less complex regulation. Based on our current knowledge, also mRNA localization appears to follow this rule. Whereas this process is found in virtually all eukaryotes, the number of proteins required to assemble a functional mRNA-transport complex appears to be considerably lower. A second important advantage is the comparably small size of yeast genomes, which can be easily manipulated and challenged by genetic screens.

For these reasons, it is not entirely surprising that the most comprehensively understood event of mRNA localization is found in the budding yeast *S. cerevisiae*.[1]

In this chapter, we will first outline our current understanding of how RNA-binding proteins contribute in different ways to mRNA localization in the budding yeast (Fig. 1A). Since mRNA localization has also been described in other fungi, we will continue with a description of mRNA localization in the human pathogen *Candida albicans* (Fig. 1B,C) and the transport of transcripts in the plant pathogen *Ustilago maydis* (Fig. 1D). A summary of RNA-binding proteins implicated in directional mRNA transport in these fungi is provided in Table 1.

Overview of mRNAs Localization in *S. cerevisiae*

During mitosis, *S. cerevisiae* undergoes unequal cell division, resulting in a larger mother cell and a smaller daughter cell.[2] Whereas the mother cell undergoes a genomic recombination in the *MAT* gene locus (from a to α or vice versa), the daughter cell remains unchanged. This genomic

*Helmholtz Zentrum München, Institute for Structural Biology and Gene Center of the Munich University, Munich, Germany.
Email: niessing@helmholtz-muenchen.de

RNA Binding Proteins, edited by Zdravko J. Lorković.

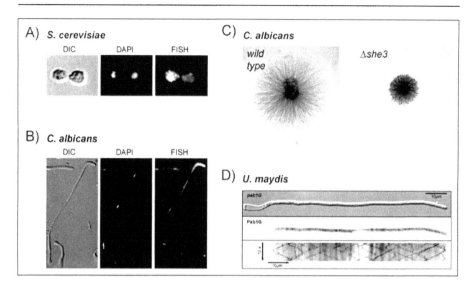

Figure 1. Images of mRNA localization in different fungi. A) In budding yeast *Saccharomyces cerevisiae* the *ASH1* mRNA is localized in the daughter cell. B) In the filamentous fungus *Candida albicans* the *ASH1* mRNA is localized to the hyphal tip by a machinery that is at least partially homologous to mRNA localization in *S. cerevisiae*. C) *C. albicans* colonies grown under cover slip. When compared to wild-type cells (left), a *she3* deletion strain (right) exhibits severe defects in filamentous growth. D) In *Ustilago maydis* mRNAs are also transported in hyphae. The upper panel shows a phase contrast image, the middle panel shows imaging of Pab1 fused to a green fluorescent proteins (Pab1G; dark spots). The lower panel contains a Kymograph, in which the positions of particles (horizontal axis; see also middle panel) are plotted against time (vertical axis). From this representation, it becomes clear that motile particles move bi-directionally and are not localized. Images in (A), (B) and (C) are used with permission from reference 45. Corresponding high resolution images were kindly provided by Sarah Elson and Alexander Johnson. Images in (D) were kindly provided by Sebastian Baumann and Michael Feldbrügge. DIC indicates differential interference contrast microscopy, DAPI nuclear staining, and FISH fluorescence in situ hybridization against *ASH1* mRNA.

recombination is mediated by the enzymatic activity of the HO endonuclease.[2] It ensures that both cells adopt different identities after cytokinesis and thus guarantees an equal distribution of mating types in a population.

About 15 years ago, it was reported that the localization of *ASH1* mRNA in the daughter cell and its local translation mediates the selective inhibition of mating-type switching in the daughter cell.[3,4] Its protein product Ash1p suppresses the HO endonuclease and therefore inhibits mating-type switching in the daughter cell.[2] A genetic screen identified the main players of this process,[5] which were referred to as SHE proteins. In the following years, work from several research groups helped to clarify their roles in mRNA localization.[1,6] Since then, more than 30 additional transcripts have been shown to be transported by the SHE machinery into the daughter cell.[7-10]

The molecular motor providing the motile activity for daughter cell localization of mRNAs is the Type V myosin She1p.[5] It is commonly referred to as Myo4p and moves its cargo along actin filaments (Fig. 2). A myosin adapter, termed She3p, directly binds with its N-terminal half to the tail of Myo4p.[11-14] Both proteins form a constitutive cytoplasmic cocomplex. In addition, the C-terminus of She3p interacts with the RNA-binding protein She2p.[11,15]

Table 1. Summary of RNA-binding proteins required for mRNA localization in yeast

Species	Protein	RNA-Binding Domain	Bound RNA Motifs	Protein Function	Defect Upon Mutation
Saccharomyces cerevisiae	She2p	Unique (Fig. 3A)	As ternary complex with She3p: to all zip-code elements and localizing mRNAs	Cotranscriptional binding of localizing mRNAs and synergistic cytoplasmic recognition of transcripts with She3p	Total loss of mRNA localization to bud tip
	She3p	Unknown	As ternary complex with She2p: to all zip-code elements and localizing mRNAs	Myosin-adapter and RNA-binding protein; acts synergistically with She2p for specific recognition of transcripts	Total loss of mRNA localization to bud tip
	Puf6p	7 Pumilio-like repeats	To UUGU motifs in localizing mRNAs	Translational repression during mRNA tranport to bud tip	Reduced mRNA localization to bud tip; premature translation
	Khd1p	3 K-homology domains	To CNN repeats in a subset of localizing mRNAs	Translational repression during mRNA tranport to bud tip	Reduced mRNA localization to bud tip; premature translation
	Loc1p	Unknown	To the E3 element of *ASH1* mRNA and to double-stranded stem loop RNAs	Unknown; (possibly required for loading of translational repressor complex)	Reduced mRNA localization to bud tip; premature translation???
Candida albicans	She3p	Unknown	Unknown	Core factor of mRNP. Inferred from homology: RNA-binding protein and myosin adapter	Loss of mRNA localization, defects in filamentous growth
Ustilago maydis	Rrm4	4 RNA-recognition motifs (RRM)	To CA-rich motifs in localizing mRNAs	Binding to mRNAs for their transport, interaction with poly-A binding protein	Loss of mRNA transport, defects in filamentous growth
	Poly-A binding protein (PAB1)	4 RNA-recognition motifs (RRM)	Inferred from homology: to poly-A tails of mRNAs	Colocalizes with Rrm4 in mRNPs Inferred from homology: Binds poly-A tails of transcripts and stimulates translation	Disruption of PAB1 interaction with Rrm4 results in loss of mRNA transport

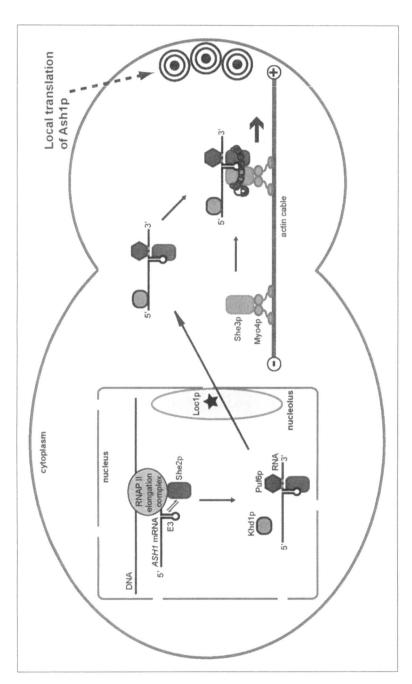

Figure 2. Schematic drawing of *ASH1* mRNA localization in the budding yeast *S. cerevisiae*. In the nucleus of the mother cell (dashed box left) *ASH1* mRNA interacts with She2p cotranscriptionally. The complex is joined by Puf6p and Khd1p, passes through the nucleolus and exports into the cytoplasm. There, a ternary interaction of She2p, She3p, and RNA achieves the stable and specific assembly of mature transport complexes. Once the complex has reached the bud tip (right), the mRNA is anchored and locally translated, resulting in local Ash1p production.

She2p is an unusual RNA-binding protein[16,17] that shuttles into the nucleus and is exported together with the *ASH1* mRNA into the cytoplasm (Fig. 2).[18-21] On its way out of the nucleus, the *ASH1* complex passes through the nucleolus.[18]

After export, the She2p-RNA complex interacts with the C-terminal half of She3p to form a larger, motor-containing complex.[11,15,22] Recent single-particle experiments with in vivo-purified transport particles demonstrated that this minimal complex, consisting of Myo4p, She3p, She2p, and RNA, indeed constitutes a motile mRNP.[23]

However, for efficient mRNA localization in vivo additional RNA-binding proteins are required (Table 1). Two of these accessory factors, termed Puf6p and Khd1p, are involved in translational repression during transport and de-repression at the site of localization (Fig. 2).[24,25] Both translational receptors seem to associate with distinct subsets of localizing mRNAs.[24-28] Like She2p, Puf6p and Khd1p shuttle into the nucleus and are most likely exported together with *ASH1* mRNA into the cytoplasm.[18,21]

The third accessory factor is Loc1p.[29] Although this protein was discovered before Puf6p and Khd1p, it remains to be the most mysterious one. Loc1p shows strictly nucleolar localization and is therefore not directly involved in cytoplasmic events (Fig. 2).[29] Nevertheless, deletion of *loc1* results in defective *ASH1* mRNA localization in the cytoplasm.[29] To date, we fail to understand why.

In addition to mRNA transport, concomitant inheritance of cortical endoplasmic reticulum (ER) by the SHE machinery has been reported.[12] For this process, Myo4p and She3p but not She2p are required. Nevertheless, She2p and *ASH1* mRNA directly associate with ER membranes.[30] It might be that this association and comigration with ER constitutes an independent and alternative path for mRNA localization. However, since we still fail to understand this interaction and its precise meaning for *ASH1* localization, aspect of ER inheritance will not be explained here further.

In the following paragraphs, we will rather discuss in more details how the above-mentioned RNA-binding proteins interact and participate to mediate directional mRNA transport. Details on the function of the myosin motor can be found elsewhere[13,14,31-33] and will not be covered by this chapter.

Bud-Localizing mRNAs

In addition to *ASH1* mRNA, She2p binding to more than 30 transcripts has been reported, of which about half clearly localize in a She2p-dependent manner.[7-10] Localizing mRNAs contain at least one cis-acting element, to which She2p binds. Such cis-acting localization elements are also known as zip-code elements[34] and usually adopt stem-loop structures.[35,36] A recent study used double fluorescence in situ hybridization (FISH) to show that transport particles contain combinations of different mRNA species that are transported together.[37]

Attempts to find common features for all of these zip-code elements failed so far, but for several well-defined zip-code elements a stem-loop structure is necessary for binding.[11,38-40] In a three-hybrid assay Pascal Chartrand and colleagues identified sequence motifs in a subset of zip-code elements that are required for She2p binding.[36] The identified motif consists of a CGA base triplet located in a loop and a single cytosine in a second loop. Both loops are separated by a double-stranded RNA helix of defined length. In a study by Joseph DeRisi and colleagues again a three-hybrid screen was used to demonstrate that this base triplet may not be sufficient for binding and that the cytosine may be dispensable.[41] It seems obvious that further functional and structural studies with different RNA-zip-code elements are necessary to understand the features of zip code RNAs that are recognized by the transport machinery.

Interestingly, it has been noted that several of the She2p-dependent mRNAs also have a functional relationship. Almost half of the She2p-dependently localizing mRNAs encode for membrane-associated proteins.[9]

The Core RNA-Binding Protein She2p

She2p has long been considered to be the only RNA-binding protein required for the specific recognition of transcripts and their incorporation into transport particles.[1,6,42] This picture only changed recently, when She3p was identified as a novel, cytoplasmic RNA-binding protein that acts in concert with She2p to recognize localizing mRNAs (see below).[22]

She2p shuttles into the nucleus with the help of the importin-alpha Srp1p.[19,21] There, She2p is located at nuclear foci, which are probably sites of *ASH1* transcription.[18] Recent chromatin-immunoprecipitation (ChIP) studies confirmed that She2p is indeed recruited to sites of active transcription, albeit with no preference for genes encoding localizing transcripts (Fig. 2).[22,43] This recruitment depends on the transcriptional elongation factors Spt4 and Spt5[43] and does not require RNA-binding.[22,43] From these findings a model has been suggested in which localizing transcripts are cotranscriptionally loaded with She2p.[43] Because it still remains unclear when and how specific She2p binds to localizing transcripts in the nucleus, additional studies will have to clarify the mechanistic details of these early events.

Once the mRNA has been synthesized, it passes through the nucleolus.[18,21] Here, the *ASH1* complex associates with the nucleolar protein Loc1p, for reasons still unknown. It is clear, however, that Loc1p does not enter the cytoplasm and thus only indirectly influences the cytoplasmic fate of *ASH1* mRNA.[29] After nuclear export, the pre-mRNP matures into an active, motor-containing particle (Fig. 2). The key interaction for this maturation is the formation of the ternary complex between She2p, She3p and localizing RNA.[22] Here, both proteins bind to the RNA and to each other. In that way, they act synergistically to achieve high binding specificity and affinity.[22] Because She3p is constitutively associated with the motor Myo4p already before this event, the ternary complex formation brings together the RNA and the motor. In summary this cytoplasmic assembly event constitutes a key quality control step for the selective transport of mRNAs with zip-code elements (Fig. 2).

Determination of the crystal structure of She2p brought the surprising finding that it consists of an unusual type of nucleic-acid binding domain.[17] Whereas the crystal structure revealed a homo-dimeric protein,[17] its analysis in solution revealed a tetrameric assembly of two dimers in a head-to-head interaction (Fig. 3A).[16,23] Disruption of She2p tetramer formation results in reduced RNA binding and impaired mRNA localization in vivo.[16] One of the most striking features of this tetramer is the four small helices that protrude at right angles from the body of the structure into the solvent (Fig. 3A). These helices are required for the interaction with She3p, for RNA binding, and thus for specific recognition of localizing RNAs.[22] The distance between two alpha-helices on one side of the structure is large enough to accommodate up to two double-stranded RNA molecules side-by-side.[16] In addition, the continuous surface between the two helices has been shown to be involved in RNA binding.[17] Thus the structural analysis indicates that the She2p tetramer is well suited for specific RNA recognition.

The more surprising it was to find that RNA binding of She2p alone has a rather low specificity.[16,18,22] This observation suggested that in vivo additional features must contribute to achieve the highly specific transport of mRNAs observed in cells. The discovery of the RNA-binding properties of She3p and its synergistic binding with She2p (see next paragraph) offered this missing piece in the puzzle.[22]

The Myosin Adapter and RNA-Binding Protein She3p

She3p has been identified by yeast-two-hybrid screens and co-immunoprecipitation experiments as a direct interactor of She2p and of the myosin motor Myo4p.[11,15,19,44] For a long time it was believed that She3p only acts as a linker between the RNA-binding protein She2p and the motor Myo4p.[1,6,31,32,42] Consistent with this assumption is that relatively tight binding was measured for the Myo4p-She3p interaction.[13] However, the interaction between She2p and She3p in absence of RNA was measured to be far too weak for a complex formation in vivo.[22] Instead, in vitro reconstitution experiments with recombinant proteins demonstrated that She3p is an RNA binding protein with rather low specificity. It has to interact with She2p to form a tight and specific

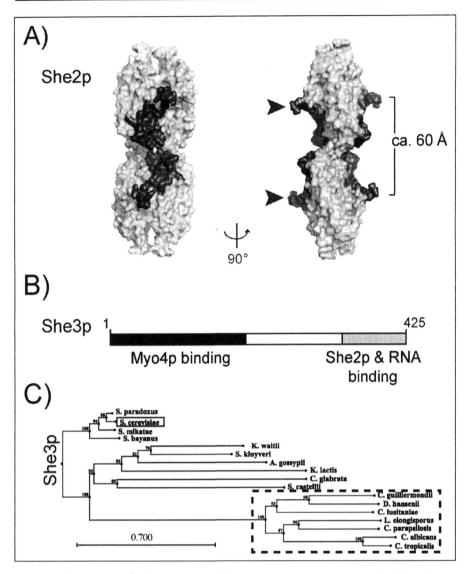

Figure 3. The core RNA-binding proteins of the *ASH1* mRNA transport complex from *S. cerevisiae*. A) Structural model of the She2p tetramer, as derived from X-ray crystallographic and small angle X-ray scattering (SAXS) analyses.[16,17] Regions highlighted in red (online version) or dark grey (print version) indicate surfaces involved in RNA binding. Two of the four protruding helices required for RNA binding and synergistic interaction with She3p are marked with arrow heads. These helices enclose at each side of the tetramer a surface region of about 60 Å in diameter that is large enough to accommodate the binding of double-stranded RNAs. B) Schematic drawing of She3p. For this protein no high-resolution structure is available. Whereas the N-terminal half of She3p interacts with Myo4p, a smaller C-terminal fragment binds to She2p and mediates synergistic binding with She2p to zip-code RNAs.[22] C) Phylogenetic tree of She3p from different yeast species with bootstrap analysis. She3p from *S. cerevisiae* is boxed. Branches to species without clear She2p homologs are enclosed by the dashed box. Numbers at nodes show bootstrap values. Images in (B) and (C) are taken with permission from reference 22.

ternary complex with localizing RNAs (Fig. 2).[22] Thus, tight She3p binding to She2p depends on the simultaneous interaction of both proteins with mRNA. This specific RNA recognition is only achieved during the late assembly of the mature transport complex in the cytoplasm (Fig. 2).

Within the 425 amino acid long protein sequence of She3p, a sub-fragment of 92 residues is sufficient for RNA binding and for synergistic interaction with She2p (Fig. 3B). Surprisingly, database searches with the protein sequence of She3p fail to yield any significant similarity to known RNA- or DNA-binding domains. It suggests that She3p might bind RNA and She2p via an unusual structural arrangement.

In certain yeast species like *C. albicans* a She3p homolog but not a She2p homolog is present in the genome (Fig. 3C).[22,45] From this, the interesting question arises whether such species also actively localize mRNAs. Indeed, *C. albicans* transports mRNAs in an She3p-dependent manner (for details, see below).[45] It suggests that *C. albicans* either uses a different She3p-binding partner than She2p or that She3p achieves higher RNA-binding specificity on its own. It will be interesting to see which of these options nature has chosen to ensure specific mRNA transport in these species.

The Translational Repressor Puf6p

Like She2p, the Puf6p protein shuttles between the cytoplasm and nucleus.[21] Puf6p binds to zip-code elements in the *ASH1* mRNA and represses translation during mRNA transport.[24,26]

Puf6p mainly consists of seven Pumilio-like repeats,[26] which are typical for Pumilio-family proteins.[46] This family of proteins usually have multiple Pumilio-like repeats that stack onto each other to produce an elongated structure.[47] These elongated, often banana-like structures bind to RNAs via their concave inner surface. Because of its pronounced homology to Pumilio-like proteins with known structures, it can be assumed that Puf6p adopts a similar shape and binds RNA in a comparable manner.

Since Puf6p could be co-immunoprecipitated with She2p from yeast extracts even after RNase digestion, a direct interaction between both proteins had been suggested.[21] Binding studies with recombinant proteins, however, failed to confirm such a direct interaction.[22] Therefore the in vitro results rather imply that *ASH1* mRNPs purified from extracts are inert enough to resist RNase treatment. Interestingly, the binding site of Puf6p on the *ASH1* mRNA is very close to a She2p binding site.[26] This vicinity of both factors could potentially help to facilitate the formation of such inert complexes.

After Puf6p has joined the nuclear complex, it shuttles with the localizing mRNAs into the cytoplasm (Fig. 2). There Puf6p acts as an associated factor while being dispensable for the formation of core motile particles.[23] In this function, Puf6p mediates the translational repression to prevent premature translation of *ASH1* mRNA during its transport to the daughter cell.[24] Upon Puf6p deletion, premature translation in the mother cell results in reduced bud-tip localization of *ASH1* mRNA and defective mating-type switching.[24,26] The mechanism of Puf6p-mediated translational repression has recently been solved: Puf6p interacts with the general translation factor eIF5B to prevent translation initiation of its bound transcripts.[24] Interestingly, this interaction is RNA-dependent. The translational repression is released by phosphorylation of Puf6p by the casein kinase II, most likely at the bud tip.[24]

The Translational Repressor Khd1p

Khd1p was identified by a systematic assessment of RNA-binding proteins to be required for efficient *ASH1* mRNA localization.[28] The genetic inactivation of Khd1 results in a reduction but not a complete loss of *ASH1* mRNA localization.[28]

This reduced localization is rather similar to the defect observed when Puf6p is inactivated.[26] Also the functions of Puf6p and Khd1p are comparable: Khd1p also acts as a translational repressor during mRNA transport, albeit by a different mechanism,[25] and both proteins shuttle between the nucleus and the cytoplasm.[18,21] Furthermore, unlike She2p but similar to Puf6p, Khd1p only binds to a single RNA region at the beginning of the open reading frame of *ASH1* mRNA.[27,28] It seems likely that Khd1p and Puf6p have redundant functions in translational repression. Alternatively, these proteins might complement each other by repressing the translation of different subsets of localizing mRNAs.

Khd1p contains three so-called K-homology (KH) domains. The KH domains are highly conserved and abundant RNA binding domains (see Chapter 1 by Michelle et al and Chapter 9 by Cléry and Allain for more details).[47,48] It suggests that Khd1p binds localizing RNAs through these domains. Khd1p also interacts with the general translation initiation factor eIF4G1. This interaction results in the inhibition of premature *ASH1* mRNA translation during its transport to the bud cell.[25] At the bud tip, Khd1p becomes phosphorylated by the membrane-associated kinase Yck1p.[25] Phosphorylation reduces the affinity of Khd1p for *ASH1* mRNA and releases the transcripts for local translation at the bud tip.

Interestingly, micro-array analysis of RNA-binding partners of Khd1p shows that this protein interacts with a large range of transcripts including only a subset of bud-localizing mRNAs.[27] A similar study for Puf6p is still missing. Further work will be required to understand the specificities and complementarity of these two translational repressors.

The Nucleolar RNA-Binding Protein Loc1p

Loc1p has been identified in a three-hybrid screen to bind to *ASH1* mRNA.[29] This protein is strictly localized in the nucleus,[29] with a sub-localization to the nucleolus.[18,49] Since the nucleolus is well-known as site of ribosome biogenesis,[50] it was surprising to discover that deletion of Loc1p results in a reduction of cytoplasmic localization of *ASH1* mRNA.[29] The second, more expected defect observed in a *loc1* mutant is impaired biogenesis of the large ribosomal subunit.[49] In *loc1Δ* cells, 35S pre-rRNA accumulates, 25S r-RNA is decreased, and 60S ribosomal subunits are not efficiently exported into the cytoplasm. Since deletion of *loc1* results in a two-fold increase of Ash1p but not of Actin levels, it has been suggested that Loc1p is involved in the regulation of *ASH1* mRNA translation.[51] An apparent uncertainty with this reasoning is that impairing ribosome biogenesis might influence translation of transcripts in an unpredictable manner and that the observed increase in Ash1p translation could still be a rather unspecific effect.

Co-immunoprecipitation experiments showed that Loc1p interacts with She2p in an RNase-independent manner.[21] This observation suggests that both proteins are part of the same protein complex or even bind directly to each other. Again, the mechanistic implications of this interaction remain to be elucidated. Attempts to identify known protein domains or even to predict the three-dimensional fold of Loc1p failed. Thus, also from a structural biology point of view, no hints can be derived on its precise molecular function.

In summary, Loc1p is involved in two seemingly unrelated, but RNA-dependent processes. It remains to be seen whether Loc1p acts more as a general factor or if it is involved in these two tasks with a very specific mission.

Long-Distance Transport of mRNAs in the Filamentous Fungus *Candida albicans*

The filamentous fungus *Candida albicans* is an opportunistic human pathogen that causes infections in immune-compromised patients.[52] It exists in different morphological forms.[53] In particular, the hyphal form has been associated with virulescence. Already a while ago, a study suggested that directional RNA transport may also take place in *C. albicans*. Ralf-Peter Jansen and colleagues had introduced *ASH1* mRNA from *C. albicans* into *S. cerevisiae* and observed its accumulation in the daughter cell.[44]

A more recent study found that *C. albicans* has a homolog of She3p but not of She2p.[45] Alexander Johnson and colleagues used a combination of genetic manipulation and biochemical experiments to show that there is indeed She3p-dependent transport of mRNAs in *C. albicans* (Fig. 4):[45] First, they observed *ASH1* mRNA localization in the tip cell of hyphae (Fig. 1B). Second, genetic deletion of both *she3* alleles resulted in loss of the tip localization of *ASH1* mRNA and impaired filamentous growth (Fig. 1C). Third, copurification of mRNAs with She3p yielded a distinct set of about

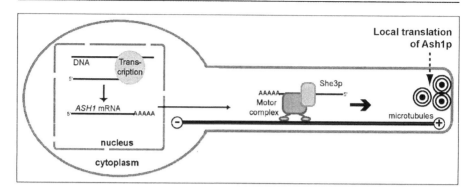

Figure 4. Schematic drawing of mRNA localization in *C. albicans*. *ASH1* mRNA and other transcripts are transported along microtubules to the hyphal tip. At the tip, Ash1p is produced by local translation. Besides *ASH1* mRNA, the RNA-binding protein and myosin adapter She3p is homologous to the mRNA localization machinery in *S. cerevisiae* (see Fig. 2). However, She2p is missing.

40 transcripts, a part of which does localize in a She3p-dependent manner. Amongst them only a fraction, including *ASH1* mRNA, have clear homologs to localizing mRNAs in *S. cerevisiae*.[45]

Obviously many questions remain regarding the identity and function of the components of the She3p-dependent transport complex in *C. albicans*. For instance, how does *C. albicans* substitute the function of She2p known from *S. cerevisiae*? Interestingly, *C. albicans* not only lacks She2p but also fails to have a Myo4p homolog.[45] In *S. cerevisiae* Myo4p is required to move *ASH1* mRNPs into the bud cell.[5] Instead, *C. albicans* only has a homolog of Myo2p,[45] which has not been reported to play a role in *ASH1* transport in *S. cerevisiae*. So another interesting question is which motor protein provides the motile activity for active transport. It will be very interesting to see how flexible evolution has handled the function of these proteins to achieve directional mRNA transport.

Long-Distance Transport of mRNAs in the Filamentous Fungus *Ustilago maydis*

Also the corn pathogen *Ustilago maydis* is a filamentous fungus that relies on active transport processes to achieve hyphal growth.[54] Like in *S. cerevisiae*, also *U. maydis* actively transports mRNAs along the cytoskeleton (Fig. 5), albeit with marked differences. In contrast to *S. cerevisiae*, not all molecular players are known yet. One central RNA-binding protein, termed Rrm4, has been

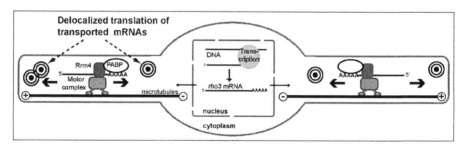

Figure 5. Schematic drawing of mRNA transport in *U. maydis*. The RNA-binding protein Rrm4 and the poly-A binding protein (PABP) are both colocalizing in transport particles. Rrm4 binds to CA-rich elements in the transported mRNAs and most likely tethers them to the motor-containing transport machinery. In contrast to mRNA localization in *S. cerevisiae* and in *C. albicans*, no anchoring and localization occurs in *U. maydis*. Instead, particles are moved in both directions, most likely distributing the RNAs instead of localizing them.

shown to be required for directional mRNA transport. Rrm4-containing particles move with high processivity to both poles.[55] Surprisingly, these particles do not become anchored at the tip, but rather turn around and move backwards in a retrograde manner. Deletion of Rrm4 results in defects in filamentous growth as well as in reduced virulence.[56]

Rrm4 contains three N-terminal RNA recognition motifs (RRMs) and one C-terminal PABC (poly-A binding protein C-terminus) domain. RRMs are highly conserved and widespread RNA-binding domains (see Chapter 1 by Michelle et al),[57] whereas the PABC domain has been shown in humans to be responsible for protein interaction with the poly-A binding protein.[58,59] Deletion of the first RRM results in a loss of function, indicating that RNA-binding is important for Rrm4 function.[55] Also mutations in the PABC domain result in defects similar to the inactivation of the entire Rrm4 protein.[55] It suggests that Rrm4 might interact with the poly-A binding protein (PABP). Indeed, the PABP of *U. maydis* (PAB1) colocalizes with Rrm4 in virtually all shuttling particles.[60]

By in vivo UV cross-linking and immunoprecipitation (CLIP) experiments,[61] about 50 transcripts were identified to interact with Rrm4.[60] These RNAs encode for proteins involved in translation, cell fate and cell polarity. Most of them contain regions with a bias towards CA-rich motifs, suggesting that Rrm4 binds preferentially to elements with a CA bias.[60]

Conclusion

Given our lack of knowledge on many molecular details of mRNA transport in *C. albicans* and *U. maydis*, it seems difficult to directly compare these species. There are, however, obvious differences. For instance the transport in *U. maydis* occurs along microtubules, whereas *ASH1* mRNA transport in *S. cerevisiae* and possibly also in *C. albicans* proceeds along actin filaments. Due to the filamentous nature of *U. maydis* and of *C. albicans* the mRNA transport occurs over much longer distances than in *S. cerevisiae*. If mRNA localization in *C. albicans* indeed proceeds along actin filaments, it would mean that the often-made assumption "short distance transport follows actin filaments—long distance transport occurs along microtubules" does not hold true in all cases. Also when we compare the RNA-binding proteins and domains that are utilized in the apparently unrelated mRNA-transport processes of *S. cerevisiae* and *U. maydis*, we find little overlap. Thus, studies in fungi suggest that there is not a single, general principle for all transport events. There might be universal principles though which remain to be discovered by future mechanistic studies.

Another very interesting issue is specificity of mRNA recognition for transport. In *U. maydis* mRNA transport appears to be less efficient and potentially also less specific than in *S. cerevisiae*. Why this difference? A likely explanation is their different functional requirements. In *S. cerevisiae* the goal is to efficiently deplete the mother cell from *ASH1* transcripts. Already mild defects in *ASH1* mRNA localization result in disrupted regulation of mating-type switching. For this reason, *ASH1* mRNA localization has to be very efficient. In the long, filamentous fungus *U. maydis*, however, active transport appears to be in place to overcome limitations in diffusion. Due to the long extensions of a cell, diffusion is not sufficient to guarantee even distribution of mRNAs and other molecules. To overcome this problem, *U. maydis* appears to use active transport.

The more details we learn about the molecular machines in these species, the better we will be able to comprehend if, despite molecular differences, there are universal principles that can be found in all of them. However, we may also find principles that only apply to distinct subsets of cellular tasks.

References

1. Müller M, Heuck A, Niessing D. Directional mRNA transport in eukaryotes: lessons from yeast. Cell Mol Life Sci 2007; 64:171-180.
2. Cosma MP. Daughter-specific repression of Saccharomyces cerevisiae HO: Ash1 is the commander. EMBO Rep 2004; 5(10):953-957.
3. Bobola N, Jansen RP, Shin TH et al. Asymmetric accumulation of Ash1p in postanaphase nuclei depends on a myosin and restricts yeast mating-type switching to mother cells. Cell 1996; 84(5):699-709.

4. Sil A, Herskowitz I. Identification of asymmetrically localized determinant, Ash1p, required for lineage-specific transcription of the yeast HO gene. Cell 1996; 84(5):711-722.
5. Jansen RP, Dowzer C, Michaelis C et al. Mother cell-specific HO expression in budding yeast depends on the unconventional myosin myo4p and other cytoplasmic proteins. Cell 1996; 84(5):687-697.
6. Paquin N, Chartrand P. Local regulation of mRNA translation: new insights from the bud. Trends Cell Biol 2008; 18(3):105-111.
7. Hogan DJ, Riordan DP, Gerber AP et al. Diverse RNA-binding proteins interact with functionally related sets of RNAs, suggesting an extensive regulatory system. PLoS Biol 2008; 6(10):e255.
8. Oeffinger M, Wei KE, Rogers R et al. Comprehensive analysis of diverse ribonucleoprotein complexes. Nat Methods 2007; 4(11):951-956.
9. Shepard KA, Gerber AP, Jambhekar A et al. Widespread cytoplasmic mRNA transport in yeast: identification of 22 bud-localized transcripts using DNA microarray analysis. Proc Natl Acad Sci U S A 2003; 100(20):11429-11434.
10. Takizawa PA, DeRisi JL, Wilhelm JE et al. Plasma membrane compartmentalization in yeast by messenger RNA transport and a septin diffusion barrier. Science 2000; 290(5490):341-344.
11. Böhl F, Kruse C, Frank A et al. She2p, a novel RNA-binding protein tethers ASH1 mRNA to the Myo4p myosin motor via She3p. EMBO J 2000; 19(20):5514-5524.
12. Estrada P, Kim J, Coleman J et al. Myo4p and She3p are required for cortical ER inheritance in Saccharomyces cerevisiae. J Cell Biol 2003; 163(6):1255-1266.
13. Heuck A, Du TG, Jellbauer S et al. Monomeric myosin V uses two binding regions for the assembly of stable translocation complexes. Proc Natl Acad Sci U S A 2007; 104(50):19778-19783.
14. Heuck A, Fetka I, Brewer DN et al. The structure of the Myo4p globular tail and its function in ASH1 mRNA localization. J Cell Biol 2010; 189(3):497-510.
15. Long RM, Gu W, Lorimer E et al. She2p is a novel RNA-binding protein that recruits the Myo4p-She3p complex to ASH1 mRNA. EMBO J 2000; 19(23):6592-6601.
16. Müller M, Richter K, Heuck A et al. Formation of She2p tetramers is required for mRNA binding, mRNP assembly, and localization. RNA 2009; 15(11):2002-2012.
17. Niessing D, Hüttelmaier S, Zenklusen D et al. She2p is a Novel RNA-Binding Protein with a Basic Helical Hairpin Motif. Cell 2004; 119:491-502.
18. Du TG, Jellbauer S, Müller M et al. Nuclear transit of the RNA-binding protein She2p is required for translational control of localized ASH1 mRNA. EMBO Rep 2008; 9:781-787.
19. Kruse C, Jaedicke A, Beaudouin J et al. Ribonucleoprotein-dependent localization of the yeast class V myosin Myo4p. J Cell Biol 2002; 159(6):971-982.
20. Powrie EA, Zenklusen D, Singer RH. A nucleoporin, Nup60p, affects the nuclear and cytoplasmic localization of ASH1 mRNA in S. cerevisiae. RNA 2011; 17(1):134-144.
21. Shen Z, Paquin N, Forget A et al. Nuclear shuttling of She2p couples ASH1 mRNA localization to its translational repression by recruiting Loc1p and Puf6p. Mol Biol Cell 2009; 20(8):2265-2275.
22. Müller M, Heym RG, Mayer A et al. A cytoplasmic complex mediates specific mRNA recognition and localization in yeast. PLoS Biol 2011; 9(4):e1000611.
23. Chung S, Takizawa PA. Multiple Myo4 motors enhance ASH1 mRNA transport in Saccharomyces cerevisiae. J Cell Biol 2010; 189(4):755-767.
24. Deng Y, Singer RH, Gu W. Translation of ASH1 mRNA is repressed by Puf6p-Fun12p/eIF5B interaction and released by CK2 phosphorylation. Genes Dev 2008; 22(8):1037-1050.
25. Paquin N, Menade M, Poirier G et al. Local activation of yeast ASH1 mRNA translation through phosphorylation of Khd1p by the casein kinase Yck1p. Mol Cell 2007; 26(6):795-809.
26. Gu W, Deng Y, Zenklusen D et al. A new yeast PUF family protein, Puf6p, represses ASH1 mRNA translation and is required for its localization. Genes Dev 2004; 18(12):1452-1465.
27. Hasegawa Y, Irie K, Gerber AP. Distinct roles for Khd1p in the localization and expression of bud-localized mRNAs in yeast. RNA 2008; 14(11):2333-2347.
28. Irie K, Tadauchi T, Takizawa PA et al. The Khd1 protein, which has three KH RNA-binding motifs, is required for proper localization of ASH1 mRNA in yeast. EMBO J 2002; 21(5):1158-1167.
29. Long RM, Gu W, Meng X et al. An exclusively nuclear RNA-binding protein affects asymmetric localization of ASH1 mRNA and Ash1p in yeast. J Cell Biol 2001; 153(2):307-318.
30. Schmid M, Jaedicke A, Du T-D et al. Coordination of endoplasmic reticulum and mRNA localization to the yeast bud. Curr Biol 2006; 16(15):1538-1543.
31. Dunn BD, Sakamoto T, Hong MS et al. Myo4p is a monomeric myosin with motility uniquely adapted to transport mRNA. J Cell Biol 2007; 178(7):1193-1206.
32. Hodges AR, Krementsova EB, Trybus KM. She3p binds to the rod of yeast myosin V and prevents it from dimerizing, forming a single-headed motor complex. J Biol Chem 2008; 283(11):6906-6914.
33. Trybus KM. Myosin V from head to tail. Cell Mol Life Sci 2008; 65(9):1378-1389.

34. Kislauskis EH, Zhu X, Singer RH. Sequences responsible for intracellular localization of beta-actin messenger RNA also affect cell phenotype. J Cell Biol 1994; 127(2):441-451.

35. Jambhekar A, Derisi JL. Cis-acting determinants of asymmetric, cytoplasmic RNA transport. RNA 2007; 13(5):625-642.

36. Olivier C, Poirier G, Gendron P et al. Identification of a conserved RNA motif essential for She2p recognition and mRNA localization to the yeast bud. Mol Cell Biol 2005; 25(11):4752-4766.

37. Lange S, Katayama Y, Schmid M et al. Simultaneous transport of different localized mRNA species revealed by live-cell imaging. Traffic 2008; 9(8):1256-1267.

38. Chartrand P, Meng XH, Hüttelmaier S et al. Asymmetric sorting of ash1p in yeast results from inhibition of translation by localization elements in the mRNA. Mol Cell 2002; 10(6):1319-1330.

39. Chartrand P, Meng XH, Singer RH et al. Structural elements required for the localization of ASH1 mRNA and of a green fluorescent protein reporter particle in vivo. Curr Biol 1999; 9(6):333-336.

40. Gonzalez I, Buonomo SB, Nasmyth K et al. ASH1 mRNA localization in yeast involves multiple secondary structural elements and Ash1 protein translation. Curr Biol 1999; 9(6):337-340.

41. Jambhekar A, McDermott K, Sorber K et al. Unbiased selection of localization elements reveals cis-acting determinants of mRNA bud localization in Saccharomyces cerevisiae. Proc Natl Acad Sci U S A 2005; 102:18005-18010.

42. Chartrand P, Singer RH, Long RM. RNP localization and transport in yeast. Annu Rev Cell Dev Biol 2001; 17:297-310.

43. Shen Z, St-Denis A, Chartrand P. Cotranscriptional recruitment of She2p by RNA pol II elongation factor Spt4-Spt5/DSIF promotes mRNA localization to the yeast bud. Genes Dev 2010; 24(17):1914-1926.

44. Münchow S, Sauter C, Jansen RP. Association of the class V myosin Myo4p with a localised messenger RNA in budding yeast depends on She proteins. J Cell Sci 1999; 112(Pt 10):1511-1518.

45. Elson SL, Noble SM, Solis NV et al. An RNA transport system in Candida albicans regulates hyphal morphology and invasive growth. PLoS Genet 2009; 5(9):e1000664.

46. Wang X, McLachlan J, Zamore PD et al. Modular recognition of RNA by a human pumilio-homology domain. Cell 2002; 110(4):501-512.

47. Auweter SD, Oberstrass FC, Allain FH. Sequence-specific binding of single-stranded RNA: is there a code for recognition? Nucleic Acids Res 2006; 34(17):4943-4959.

48. Messias AC, Sattler M. Structural basis of single-stranded RNA recognition. Acc Chem Res 2004; 37(5):279-287.

49. Urbinati CR, Gonsalvez GB, Aris JP et al. Loc1p is required for efficient assembly and nuclear export of the 60S ribosomal subunit. Mol Genet Genomics 2006; 276(4):369-377.

50. Kressler D, Hurt E, Bassler J. Driving ribosome assembly. Biochim Biophys Acta 2010; 1803(6):673-683.

51. Komili S, Farny NG, Roth FP et al. Functional specificity among ribosomal proteins regulates gene expression. Cell 2007; 131(3):557-571.

52. Hernday AD, Noble SM, Mitrovich QM et al. Genetics and molecular biology in Candida albicans. Methods Enzymol 2010; 470:737-758.

53. Sudbery P, Gow N, Berman J. The distinct morphogenic states of Candida albicans. Trends Microbiol 2004; 12(7):317-324.

54. Zarnack K, Feldbrügge M. Microtubule-dependent mRNA transport in fungi. Eukaryot Cell 2010; 9(7):982-990.

55. Becht P, König J, Feldbrügge M. The RNA-binding protein Rrm4 is essential for polarity in Ustilago maydis and shuttles along microtubules. J Cell Sci 2006; 119(Pt 23):4964-4973.

56. Becht P, Vollmeister E, Feldbrügge M. Role for RNA-binding proteins implicated in pathogenic development of Ustilago maydis. Eukaryot Cell 2005; 4(1):121-133.

57. Stefl R, Skrisovska L, Allain FH. RNA sequence- and shape-dependent recognition by proteins in the ribonucleoprotein particle. EMBO Rep 2005; 6(1):33-38.

58. Kozlov G, De Crescenzo G, Lim NS et al. Structural basis of ligand recognition by PABC, a highly specific peptide-binding domain found in poly(A)-binding protein and a HECT ubiquitin ligase. EMBO J 2004; 23(2):272-281.

59. Kozlov G, Trempe JF, Khaleghpour K et al. Structure and function of the C-terminal PABC domain of human poly(A)-binding protein. Proc Natl Acad Sci U S A 2001; 98(8):4409-4413.

60. König J, Baumann S, Koepke J et al. The fungal RNA-binding protein Rrm4 mediates long-distance transport of ubi1 and rho3 mRNAs. EMBO J 2009; 28(13):1855-1866.

61. Ule J, Jensen K, Mele A et al. CLIP: a method for identifying protein-RNA interaction sites in living cells. Methods 2005; 37(4):376-386.

CHAPTER 6

mRNA Localization in Plants and the Role of RNA Binding Proteins

Kelly A. Doroshenk,[1] Andrew J. Crofts,[2] Haruhiko Washida,[3]
Mio Satoh-Cruz,[1] Naoko Crofts,[2] Yongil Yang,[1] Robert T. Morris,[4]
Thomas W. Okita,*[1] Masako Fukuda,[5] Toshihiro Kumamaru[5]
and Hikaru Satoh[5]

Abstract

Intracellular mRNA localization is an important regulatory aspect of eukaryotic gene expression. Targeted, active transport requires RNA cis-localization elements, trans-acting RNA binding proteins (RBPs) that recognize these sequences and an intact cytoskeleton. Although highly prevalent and well-studied in metazoans and yeast, relatively few reports of RNA localization exist for plants. The asymmetric sorting of storage protein mRNAs in developing rice endosperm, however, has proven to be an excellent model for study. Proper storage protein synthesis and deposition into separate subcellular compartments requires direct targeting of the corresponding mRNA to distinct subdomains of the cortical endoplasmic reticulum (ER). Although evidence supports the existence of two regulated RNA localization pathways to the ER that are dependent upon RNA cis-localization sequences and cytoskeleton, little is known about the trans-acting factors required for transport. Much remains to be discovered about plant RBPs in general and the identification and subsequent characterization of cytoskeleton-associated RBPs and accessory proteins involved in seed storage protein RNA localization will further our knowledge of cytoplasmic gene expression events.

Introduction

RNA binding proteins (RBPs) are involved in every aspect of eukaryotic gene expression, from transcription and pre-mRNA processing within the nucleus to transport, stability, translation and degradation in the cytoplasm.[1,2] The functional diversity and variety of target ligands characteristic of RBPs can be attributed to the occurrence of one or more RNA binding domains within the protein, such as the RNA recognition motif (RRM) or K-homology (KH) domain.[3] The identification of these conserved domains has proven to be a useful tool in the prediction of putative RBPs, particularly in plants, which have garnered less experimental scrutiny than their counterparts in metazoans and yeast.[4] In *Arabidopsis*, over 200 RRM and KH domain-containing proteins alone have been identified, many of which have not been functionally characterized.[5] The actual number of proteins with RNA binding or modification activity, however, is likely to

[1]Institute of Biological Chemistry, Washington State University, Pullman, Washington, USA;
[2]International Liberal Arts Program, Akita International University, Akita, Japan; [3]Laboratory
of Plant Molecular Genetics, Nara Institute of Science and Technology, Ikoma, Nara, Japan;
[4]School of Molecular Biosciences, Washington State University, Pullman, Washington, USA;
[5]Faculty of Agriculture, Kyushu University, Fukuoka, Japan.
*Corresponding Author: Thomas W. Okita—Email: okita@wsu.edu

RNA Binding Proteins, edited by Zdravko J. Lorković.
©2012 Landes Bioscience.

be much higher. Assigning putative functions based on the presence of RNA binding domains has its limitation, as many RBPs are plant-specific and therefore lack homologues in the more well-studied models.[5,6] Given the increasingly apparent importance of RBPs in photosynthetic organisms, especially in aspects of signaling and development,[4,6] it is clear much work is necessary to understand their involvement in nuclear and cytoplasmic gene expression.

Cytoplasmic RBPs with a variety of proposed roles in RNA stability, storage, transport, translation and degradation have been experimentally identified using both biochemical and genetic techniques (recently reviewed in ref. 7; for a review including nuclear localized RBPs, please see refs. 5-6). RBPs are likely to play a particularly important role in the targeted localization of RNA within the cell. Although our knowledge of this process and the factors involved is particularly limited in plants, studies of other eukaryotes have shed light on the importance of RNA localization in organismal development (see Chapter 4 by Quattrone et al and Chapter 5 by Niessing). In this chapter, we will discuss what is known about RNA sorting in plants and how it is utilized, with a particular focus on the efforts of our laboratory over the past 20 years using developing rice seed as a model. As RNA localization is likely equally prevalent in photosynthetic organisms as it is in metazoans and yeast, current and future work on the identification and characterization of cytoplasmic RBPs and other accessory proteins involved will advance our general understanding of the mechanisms and impact of this form of gene regulation in plant development.

RNA Localization—A Brief Introduction

The intracellular transport of mRNAs is a common phenomenon in model eukaryotes, from yeast and animals to algae and higher plants.[8-10] RNA localization to specific sites within the cell not only allows for spatial and temporal control of gene expression, but also enhances the efficiency of protein synthesis and avoids deleterious protein interactions.[9] The well-studied RNA localization pathways in yeast and animals serve a multitude of biological purposes including the establishment of cell polarity, determination of cell fate, targeting of proteins to particular organelles and formation of functional protein-protein interactions between locally translated proteins.[11,12] It is now clear that RNA localization is a ubiquitous process, as it was recently demonstrated in *Drosophila* that 71% of the nearly 3400 transcripts analyzed exhibited targeted transport within the cell, resulting in asymmetric protein accumulation.[13]

Mechanisms of RNA localization include localized synthesis of transcripts, passive diffusion and subsequent trapping and anchoring at the proper destination site, localized degradation and/or stabilization and active transport along the cytoskeleton.[11,14,15] The presence of multiple cis-acting localization elements, or zipcodes,[16] within the RNA sequence mediates transport and may serve to promote efficient recruitment of trans-acting factors by a structurally based mechanism.[9] Most examples of RNA localization appear to occur via active transport. As first proposed by Wilhelm and Vale,[17] active transport of localized RNAs involves three steps: Assembly of a large, cytoplasmic, ribonucleoprotein transport particle consisting of trans-acting RBPs that recognize cis-acting localization sequences within the target mRNA; transport of the particle by motor proteins along microtubules or microfilaments; and final anchoring at the destination site. Increasing evidence now suggests the fate of localized mRNAs is initially decided by events within the nucleus, not the cytosol.[18] Here, processing of pre-mRNAs into mature transcripts competent for nuclear export involves the recruitment of factors that assemble to form a ribonucleoprotein complex (RNP). Both nuclear and cytosolic RNPs are highly dynamic and undergo frequent remodeling, although some components recruited in the nucleus have been found to associate with mRNAs at the site of translation.[1,2,19] This includes some members of the heterogeneous nuclear ribonucleoprotein (hnRNP) family, which can have multiple functions in RNA splicing, stability, localization and translation.[2,20] The combination of dual localization and multifunctionality may serve as a form of regulation linking gene expression events within the cell.[1,2,4]

Cytosolic RNP transport particles involved in localization are relatively large and can contain multiple RNAs and RBPs as well as accessory proteins such as translation factors.[9,11] RNAs are transported in a translationally arrested state until final anchoring at the target destination, where protein synthesis resumes.[21] Movement within the cell is dependent upon microtubules and/or actin filaments and their associated motor proteins, although the nature of the cytoskeleton-RNP interaction is not clearly understood.[22-24] Interestingly, recent studies demonstrate functionally related mRNAs colocalize to the same region of the cell and individual RBPs interact with distinct, but related, sets of RNAs.[25] These results suggest some cis- and trans-factors necessary for transport are shared, but the prevalence of cotransport within the same RNP remains to be seen. A number of RNA cis-localization elements have been described in yeast and animals, as well as the interacting RBPs necessary for transport.[2,9,18,26] A single zipcode sequence can be recognized by multiple RBPs, as is the case for *Xenopus* vg1 mRNA,[27,28] or by a complex of proteins, as with *Drosophila* bicoid,[29] highlighting the complexity underlying the regulation of cytosolic gene expression in eukaryotes.

RNA Localization in Plants

While great strides have been made regarding the elucidation of RNA targeting in metazoans and yeast, much less is known about this process in photosynthetic organisms. The relatively few studies that have been done indicate RNA transport occurs both between and within cells. In higher plants, intercellular trafficking via plasmodesmata is supported by the presence of both coding and noncoding RNAs and RBPs in phloem sap and is thought to serve as a signaling mechanism in response to stress and in coordinating gene expression during plant development.[30-33] Within algal and plant cells, RNA localization may play a role in cell fate determination, polarity and embryo development.[34-37] RNA targeting may also promote localized protein synthesis at specific organelles, including chloroplasts, mitochondria and endoplasmic reticulum (ER).[24,38-41]

Perhaps one of the best studied examples of RNA localization to the ER is the transport of storage protein mRNAs in rice. Rice is unique among plants in that it accumulates large amounts of not one, but two classes of storage proteins within the endosperm: Prolamines, typical of most cereals, and glutelins, homologous to the 11S globulins synthesized by legumes.[42] It also accumulates a third minor-type, α-globulins. Storage proteins constitute up to 90% of the total protein in rice seed and provide a reserve of carbon and nitrogen for the developing seedling. Interestingly, the different types of storage proteins do not share a common site of deposition in the cell, but instead are sorted to separate membrane-bound compartments. After translation and translocation into the ER, prolamine polypeptides accumulate as intracisternal inclusion granules within the lumen, forming spherical protein bodies (PBs), while globulins and glutelins are sorted to large, irregularly shaped protein storage vacuoles (PSVs) via the Golgi.[43-45] Prolamine PBs are largely concentrated in the cortical region of the cell, where they are closely associated with the cytoskeleton.[46] In fact, this region is highly enriched in actin microfilaments and translation factors and serves as the major site of protein synthesis in endosperm cells.[10,46,47] This synthesis occurs not only on the rough ER that delimits the prolamine PBs (PB-ER), but also on a network of lamellar and tubular membranes collectively known as cisternal-ER (cis-ER),[44,48] which together with the PB-ER forms the cortical ER (Fig. 1).

How does the asymmetric accumulation of storage proteins occur? Our laboratory has spent many years actively addressing this question and it is now clear that targeted transport of storage protein mRNAs to different subdomains of the cortical ER is responsible for subsequent protein localization. Using a combination of biochemical and high resolution in situ hybridization techniques, prolamine mRNAs were found predominately on PB-ER, while glutelin RNAs were enriched on cis-ER.[48,49] Further studies revealed prolamine mRNAs exist as relatively large particles in the cytosol that move in an actin-dependent, stop and go manner en route to the prolamine PBs of the cortical ER.[50] This movement was also dependent upon functional initiation of translation suggesting that the RNA is transported as a translationally arrested complex.[49] Localization of prolamine RNA requires two cis-localization sequences, one each within the 5′ coding sequence and 3′ untranslated region (UTR).[51] Deletion of one or both of these sequences resulted in partial or complete mislocalization, respectively, of RNA from PB-ER to cis-ER. Moreover, the localization of reporter RNAs to the cis-ER was

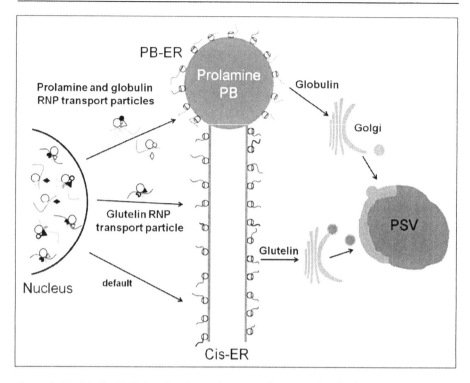

Figure 1. Model of mRNA localization pathways to the cortical endoplasmic reticulum (ER) in developing rice endosperm. After transcription, ribonucleoprotein (RNP) complexes assembled in the nucleus are exported to the cytosol. Here, remodeled RNP transport particles containing storage protein mRNAs are actively targeted to the cortical ER via the cytoskeleton by three pathways. Upon reaching protein body ER (PB-ER), prolamine and globulin mRNAs are translated and translocated into the lumen, where prolamine polypeptides form intracisternal granules and accumulate. Globulins are transported to protein storage vacuoles (PSVs) in dense vesicles via the Golgi. Glutelin mRNAs are targeted to adjacent cisternal ER (cis-ER) and the subsequent protein is sorted to PSVs. A third default localization pathway is inferred for mRNAs encoding nonsecretory proteins. Regulated transport of storage protein mRNAs to distinct subdomains of the ER is directly responsible for proper protein synthesis and sorting, and in some instances, avoids deleterious protein-protein interactions within the lumen. Reproduced with permission from Crofts AJ et al. Biochem Cell Biol 2005; 83:728-737;[53] ©2008 NRC Research Press.

disrupted by addition of the prolamine gene sequence, resulting in transport to the PB-ER. Interestingly, addition of glutelin 3′ RNA sequences redirects RNAs targeted to the PB-ER to the cis-ER, indicating glutelin RNAs contain at least one zipcode necessary for proper sorting to the cis-ER.[49,51] Later studies identified multiple glutelin cis-localization sequences in the 5′ and 3′ coding sequences, as well as 3′ UTR.[52] Together, these results suggest there are at least two regulated RNA transport pathways to the cortical ER in rice endosperm: One to the PB-ER and one to the cis-ER, with the latter exhibiting dominance over the former (Fig. 1).[49,51] The transport of reporter RNAs and prolamine RNAs lacking both zipcodes to the cis-ER supports the existence of a third default pathway as well.[49,51]

Characterization of rice mutants defective in storage protein synthesis, sorting or processing has provided further evidence for multiple RNA localization pathways in seed.[24,53] One class of mutants has been identified based on the abnormal accumulation of the 57 kD glutelin precursor, which is normally proteolytically cleaved into acidic and basic subunits in PSVs.[54] Interestingly, in situ reverse transcription PCR (RT-PCR) analysis revealed some of these mutants exhibit storage protein RNA mislocalization. In *glup2*, prolamine mRNA is mistargeted to the cis-ER, while in the non-allelic *glup4* and *glup6* mutants, glutelin mRNA is partially mislocalized to the PB-ER (ref. 55, Washida et al, unpublished results). The latter finding suggests the two regulated prolamine and glutelin RNA transport pathways may not be entirely unrelated. *glup5* is a globulin RNA mistargeting mutant. Although globulin proteins are transported to and accumulate in the periphery of glutelin-containing PSVs,[44] globulin mRNAs are targeted to PB-ER (Washida et al, unpublished results). In *glup5*, however, globulin RNAs are mis-sorted to the cis-ER (Washida et al, unpublished results). As prolamine RNA transport to the PB-ER is unaffected in *glup5*, it is possible more than one mechanism of localization exists to this region of the cortical ER, adding another layer of complexity. Subsequent studies of the *glup4* and *glup6* mutants revealed the importance of RNA localization on proper protein sorting. Partial mislocalization of glutelin mRNAs to PB-ER in these mutants leads to abnormal protein accumulation in both PSVs and prolamine PBs, as well as the formation of large dilated membranous structures (Fukuda et al, unpublished results). The targeting of storage protein RNAs to separate subdomains of the ER may serve to avoid deleterious interactions between characteristically different proteins within the ER lumen.[24,53] It may also function to locally concentrate proteins and thus promote proper formation of intracisternal granules that eventually develop into protein bodies.

The relationship between RNA and protein targeting has been demonstrated in other plants as well. Earlier biochemical[56] and in situ RNA localization[57] studies concluded that RNAs that coded for the maize prolamines (zeins) were randomly distributed on PB-ER and cis-ER. Using a more sensitive and facile in situ RT-PCR analysis, however, RNAs encoding maize zeins and 11S globulins (legumins-1) were found to be asymmetrically localized to ER surrounding zein protein bodies or adjacent cisternal ER, respectively, in endosperm cells.[58] When heterologously expressed in rice, zein mRNAs are transported to PB-ER and the subsequent protein accumulates within prolamine PBs.[51] Similar to rice prolamine, cis-localization sequences are required for targeting zein mRNA to the PB-ER.[59] When the RNA is redirected to cis-ER the protein is instead sorted to the PSV.[59] Improper protein deposition was also evident when sunflower 2S albumin mRNA was mistargeted to the PB-ER. RNA targeting to specific ER subdomains may thus be the underlying basis for the use of different endomembrane sites as storage compartments among seed plants.

It is clear that proper sorting of storage proteins in rice is dependent upon the targeted transport of mRNAs to separate subdomains of the ER. As described above, deletion studies and analysis of genetic mutants suggest there are three RNA localization pathways from the nucleus to the cortical ER (Fig. 1). Whether the transport particles involved are distinct for each RNA species or share common components is not yet known. Studies of the *glup* mutants, however, suggest the glutelin and prolamine RNA transport pathways are interrelated, as glutelin mRNA is only partially mislocalized to the PB-ER in *glup4* and *glup6* (ref. 55, Washida et al, unpublished results). Considering the available evidence, one possible model of RNA localization in rice is that RNAs encoding storage proteins and nonsecretory proteins are exported from the nucleus within a single particle, which then splits into three separate pathways en route to the cortical ER (Fig. 2).[24,53] These pathways are hierarchal in nature, with the glutelin RNA pathway to the cis-ER dominant over the prolamine RNA pathway to the PB-ER, which in turn is dominant over the default constitutive pathway to the cis-ER. While some components may be shared, it is also likely that certain factors provide specificity at the branching point to direct regulated transport to either PB-ER or cis-ER. These factors include cis-localization sequences as well as interacting RBPs and other accessory proteins, suggesting that plant RNP transport particles, like those studied in other eukaryotes, are highly dynamic.

The targeting of rice storage protein mRNAs to the cortical ER requires cis-localization sequences that are likely recognized by trans-acting RBPs. In plants, relatively little is known about cytoskeleton-associated RBPs that may be involved in mRNA localization as only a few cases of polarized transport have been reported.[10,24,53] Some examples of identified RBPs are listed in Table 1. Proteomic analysis of the tubulin binding fraction in *Arabidopsis* identified 26 RBPs,

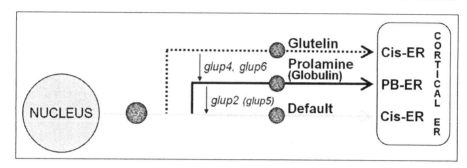

Figure 2. The relationship between rice storage protein RNA localization pathways to the cortical ER. Genetic and biochemical evidence suggests the glutelin and prolamine RNA transport pathways to the cis-ER and PB-ER, respectively, are inter-related yet hierarchal in nature. The glutelin mRNA localization pathway is dominant over the prolamine RNA transport pathway, which in turn is dominant over the constitutive pathway to the cis-ER. One possibility for this is after export from the nucleus, common RNP transport particle components are utilized to target RNAs to the cortical ER. En route, specific factors are recruited, perhaps mediated by RNA cis-localization sequences, and the pathway branches to the specific ER subdomains. The effect of each *glup* mutant on RNA localization is shown. Globulin mRNA is shown in parenthesis, as less is known about its targeting to PB-ER. At least one factor is required for transport specificity, though, as the *glup5* mutant exhibits globulin RNA mislocalization to the cis-ER. Since prolamine RNA transport remains unaffected, the possibility of more than one localization pathway to the PB-ER exists. Adapted from Crofts et al.[53]

though many of those have not been functionally characterized.[60] A few reports, however, do describe the characterization of proteins that may be involved in cytoplasmic RNA transport. In rice, a peroxisomal multifunctional protein (MFP) involved in fatty acid biosynthesis was found not only to bind microtubules, but also RNA, suggesting a possible role in localization or translation.[61,62] In *Arabidopsis*, members of the Puf-like family of proteins, involved in mRNA decay and translational regulation in animals, exhibit RNA binding activity and partially localize to particles within the cytoplasm that move in an actin-dependent manner.[63] These results could indicate a similar role for these proteins in plants as in animals or perhaps in RNA transport. Also in *Arabidopsis*, plant-specific RBP PHIP1 was found to bind Ran2 mRNA and its protein product as well as other GTPases involved in cell plate formation.[64] Furthermore, PHIP1 localizes to the cell plate, suggesting a role in targeting RNAs of cell plate biosynthetic enzymes to this region, a process the authors hypothesize could occur by vesicle trafficking along microtubules. An hnRNP in Chinese pine that colocalizes with polysomes and microtubules in the cytoplasm may serve to traffic mRNAs to the cortical region of fertilized egg cells in the establishment of cell polarity.[65] Finally, in maize, an RBP expressed only in seed was found in the cytoplasmic fraction and bound zein mRNA in vitro, though it remains to be seen whether it is involved in storage protein gene expression in vivo.[66]

It is obvious that much work still needs to be done regarding the identification and characterization of RBPs and other ancillary factors necessary for RNA localization in plants. The following sections highlight the efforts of our laboratory to contribute to the understanding of this field and ultimately to elucidate the mechanisms underlying cytoplasmic gene expression.

Identification of Cytoskeleton-Associated Plant RBPs Involved in RNA Localization

Our lab has sought to elucidate the mechanism of RNA localization in plants through the identification of trans-acting factors required for storage protein mRNA transport. As early studies indicated the involvement of the cytoskeleton in the transport and translation of rice prolamine mRNAs associated with PB-ER,[46,67] efforts have largely been directed towards identifying proteins associated with a cytoskeleton-protein body enriched fraction from seed

Table 1. Studies describing experimentally identified plant RBPs associated with the cytoskeleton and/or RNA localization

Organism	Protein(s) Identified	Summary	References
A. thaliana	Tubulin-associated RBPs	26 RBPs identified including TudorSN, PolyA binding protein, RNA helicases, and proteins of unknown function containing RNA binding domains	60
A. thaliana	PHIP1	Interacts with Ran2 mRNA; May be involved in localization of RNAs encoding proteins involved in cell plate formation via the cytoskeleton during cytokinesis	64
O. sativa	Puf family of proteins	Studied members exhibit RNA binding activity; Some localize to cytoplasmic particles whose movement is dependent on intact actin filaments	63
O. sativa	PolyA binding protein	Identified from cytoskeleton-protein body enriched seed extract	68
O. sativa	OsTudorSN	Cytoskeleton associated; Binds storage protein mRNAs and colocalizes with RNA movement particles; RNAi seeds exhibit reduced storage protein gene expression	70, 78
O. sativa	Peroxisomal multifunctional protein (MFP)	Localizes to cortical microtubules; Exhibits RNA binding activity	61, 62
O. sativa	Poly(U) binding fraction	148 cytoskeleton-associated, nucleic acid binding proteins identified, including 20 putative RBPs	79
O. sativa	Prolamine mRNA binding proteins	18 cytoskeleton-associated RBPs identified as prolamine mRNA zipcode binding proteins under stringent binding conditions, including putative hnRNPs 77 cytoskeleton-associated, putative RBPs identified as prolamine mRNA zipcode binding proteins under low stringency binding conditions	82 87
P. tabulaeformis	hnRNP	Localizes to nucleus and microtubule-associated polysomes in polarizing egg cells post-fertilization	65

using proteomic, affinity chromatography, and UV-crosslinking approaches.[68-70] OsTudorSN, previously called Rp120, is one protein that was identified from these studies. OsTudorSN is an ortholog of the mammalian p100 protein which possesses multiple functions. It was initially characterized as a transcriptional co-activator and later found to be a component of the RNA-induced silencing complex (RISC) and associated with the major nuclease activity of hyper-edited RNAs.[71-75] More recently, Arabidopsis TudorSN was found to be highly expressed in seeds and have a possible role in seedling development.[76] In addition, a cytoplasmic localization for Arabidopsis TudorSN was reported with some populations colocalizing with ER.[77] Further analysis revealed TudorSN may be involved in stress response, possibly through the stabilization of RNAs.[77] In rice seed, OsTudorSN is a cytoplasmic-localized, cytoskeleton-associated protein that exists in an RNase sensitive, high molecular weight complex and binds both prolamine and glutelin RNAs.[69,70,78] Immunofluorescence microscopy shows OsTudorSN colocalizes with both PB-ER and cisternal ER, as well as to prolamine mRNA transport particles. Indeed, GFP-tagged OsTudorSN is seen as particles whose movement is dependent upon intact actin filaments.[78] Developing rice seeds from OsTudorSN RNAi transgenic plants not only express less OsTudorSN transcript and protein, but also less prolamine RNA and protein. Together, these results suggest a role in storage protein mRNA gene expression. Current efforts include identifying other RNA targets of OsTudorSN to further characterize its function in seed.

With the advance of proteomic technology, a more global approach to identify cytoskeleton-associated RBPs in rice has been taken.[79] Using Poly(U)-Sepharose affinity chromatography, nucleic acid binding proteins from a cytoskeleton-enriched seed extract were isolated and analyzed by two-dimensional gel electrophoresis and mass spectrometry. 148 unique proteins were identified, including OsTudorSN, putative RBPs, translation factors and metabolic enzymes, an indication of the variety of activities associated with the cytoskeleton. Among the 20 proteins found with possible roles in RNA processing (not including translation factors), six had RNA binding domains characteristic of heterogeneous nuclear ribonucleoproteins (hnRNPs). Of the four proteins containing RNA recognition motifs, two demonstrated homology to a pumpkin polypyrimidine tract binding protein (PTB) that binds RNA and is a component of mobile RNP complexes in phloem sap.[33] In animals, PTBs have been found to be part of RNP complexes involved in RNA localization.[80,81] Two other hnRNP-like proteins contained KH domains, which in metazoans are found in proteins that play a role in transcription, RNA localization and translational regulation (see Chapter 1 by Michelle et al, Chapter 5 by Niessing and Chapter 9 by Cléry and Allain for more details); however in plants, little is known about their function.[5] Interestingly, of those proteins identified in the Poly(U) binding fraction with potential roles in RNA metabolism, only 4 were previously described in other studies, highlighting the validity of such an approach as well as the need for continued research in this area. A few RBPs of particular interest identified in this survey have been selected for further study to assess their possible role in plant cytoplasmic gene expression.

As our lab is particularly interested in storage protein mRNA localization, a more targeted approach to identify prolamine mRNA binding proteins was undertaken using a biotinylated prolamine zipcode sequence as bait. In the presence of a high concentration of the competitive binding inhibitor heparin, Crofts et al[82] identified 18 proteins from a cytoskeleton-enriched developing rice seed extract that interacted with prolamine zipcode RNA, although 3 proteins were found to bind equally well to a nonzipcode control. Of the proteins identified in this study, only one was also found in the proteomic analysis of the Poly(U) binding fraction from rice seed.[79] This may be due to a low level of protein expression that was overcome by subsequent enrichment using a specific RNA sequence as bait. Included in the prolamine zipcode capture findings were 7 proteins with homology to the hnRNP family, as well as a ribosomal, glycine rich and cold shock domain protein. Both glycine rich and cold shock domain proteins have been implicated in a wide variety of RNA processing events in plant development and stress response (see Chapter 8 by Kang and Kwak).[6,83] In Drosophila, cold shock domain protein Yps was found as part of a RNP transport complex involved in oskar mRNA localization to the posterior pole of oocytes.[84] Two putative oligouridylate binding proteins, RBP-L and RBP-P, were also found to interact with prolamine zipcode RNA.[82] RBP-L is similar to RBP45 and RBP47, proteins identified from *Arabidopsis* and *Nicotiana plumbaginifolia* that are suggested to play a role in nuclear pre-mRNA processing and stability in tobacco.[85] Interestingly, RBP47 is also found within cytoplasmic RNP complexes in response to stress, which may indicate possible multifunctionality.[86] To confirm the interaction with prolamine mRNA and assess subcellular localization, antibodies were generated to many of the identified prolamine RNA binding proteins.[82] Most showed some level of specificity to the prolamine zipcode sequence, including RBP-A, a putative hnRNP. Using immunofluorescence microscopy, RBP-A was found not only localized to the nucleus, but also the cortical cisternal ER and microtubule network. In addition, immunoprecipitation and subsequent RT-PCR analysis demonstrated that both prolamine and glutelin RNAs associate with RBP-A.[82] Together, these results suggest a possible role for RBP-A in nuclear RNA processing events, RNA sorting, anchoring to the ER and/or microtubules or translational regulation. Studies are currently ongoing for the other RBPs identified (discussed in more detail below).

A second, similar study was conducted utilizing biotinylated prolamine 5′ coding sequence zipcode RNA as bait but with less stringent binding conditions (low salt, no heparin) in an attempt to capture a larger population of cytoplasmic-localized, cytoskeleton-associated RBPs.[87] For instance, although *Os*TudorSN binds storage protein mRNAs and likely plays a role in their localization,[78] it also binds heparin[70] and therefore was not detected in the initial prolamine capture experiment.[82] Using less stringent binding conditions, 132 putative RBPs were identified that interacted with the prolamine zipcode sequence, including 12 of the 18 RBPs from the first prolamine zipcode capture experiment[82] and *Os*TudorSN. Of interest are the 77 additional RBPs that were not previously found in the Poly(U) binding fraction[79] or stringent prolamine zipcode capture[82] experiments, including a number of uncharacterized proteins with predicted RNA binding domains. Collectively, these studies have yielded over 200 putative RBPs and the results have been compiled in the RiceRBP database (http://www.bioinformatics2.wsu.edu/RiceRBP), which contains detailed information for each RBP such as cDNA and protein sequences, predicted domains and transcript expression, as well as methodology and mass spectrometry data.[87] In addition, RiceRBP can be used to look at phylogenetic relationships among the rice RBPs as well as to search for orthologs in other species. This will prove useful for other studies of RBPs in plants, as a comparison to the POGS/PlantRBP database[88] revealed only 37% of the RiceRBP entries were actually predicted to be RBPs based on sequence similarity, emphasizing the importance of such an experimentally verified database.[87] Future updates include RNA-protein and protein-protein interactions as they become available to further our understanding of this functionally diverse class of proteins in plants.

Characterization of RBPs Involved in RNA Localization

Much effort has been directed toward the identification of proteins that interact with rice storage protein mRNAs in order to understand the mechanism of RNA transport to the cortical ER. Follow-up functional studies are now necessary to determine what role these proteins may play in plant gene expression. Initial characterization of RBP-A, a prolamine zipcode RNA binding protein, demonstrates this putative hnRNP binds both prolamine and glutelin RNA and colocalizes to the nucleus, microtubules and cortical cis-ER.[82] The dual localization of RBP-A suggests that it may have multiple functions, shuttling between the nucleus and cytoplasm. Indeed, protein components of nuclear-assembled RNP complexes have been found to have different roles depending on their localization within the cell.[18] Based on its localization, RBP-A may be involved in pre-mRNA processing within the nucleus, anchoring at the ER and/or on microtubules, or translational regulation. It is likely RBP-A is not specific only to storage protein gene expression, as microarray analysis of RNAs associated with RBP-A reveals a set of targets that are highly expressed (Crofts et al, unpublished results). It would be of interest to determine whether the RNAs that RBP-A interacts with are the same in both the nucleus and cytoplasm, an indication of spatial-dependent functionality.

Two other putative hnRNP proteins identified in the prolamine zipcode capture experiments are RBP-D and -I.[82] Similar to RBP-A, RBP-D exists as multiple populations within the cell, colocalizing with the nucleus as well as particulate structures associated with actin filaments that may represent RNA transport particles (Crofts et al, unpublished results). RBP-D also binds both prolamine and glutelin RNAs (Crofts et al, unpublished results). Together, RBP-A, -D and -I all contain two RRM domains and share significant homology with HRP48 and Squid, two hnRNP A/B homologs in *Drosophila*. HRP48 and Squid are required for proper oskar and gurken mRNA localization, respectively, whose gene products are important for the development of axis formation in oocytes.[89-92] Evidence suggests a role in translational regulation for both HRP48 and Squid[89,91,93] and in addition to the cytoplasm, each has been found to localize to the nucleus.[94,95] Within the nucleus, HRP48 also has an alternate function in splicing.[96] Interestingly, further studies indicate Squid and HRP48 are involved in the transport of both oskar and gurken mRNAs and that the two proteins interact in vivo.[93,97] To assess whether similar hnRNP complexes exist in rice, co-immunoprecipitation (coIP) experiments have been done using antibodies to RBP-A and RBP-I followed by subsequent mass spectrometry and immunoblot analysis of the bound protein fraction. Initial results suggest RBP-A interacts with RBP-I, yet RBP-I associates with both RBP-A and RBP-D (Yang and Okita, unpublished results). It remains to be seen whether these interactions are direct, indirect, or RNA dependent. Interestingly, both glyceraldehyde 3-phosphate dehydrogenase (GAPDH) and sorbitol dehydrogenase (SDH) were also found in RBP-A and RBP-I coIP results. Although GAPDH is traditionally thought of as being involved in carbohydrate metabolism, it is also a well-documented RNA binding protein with putative roles in gene expression, RNA transport and translation.[98] Indeed, a number of metabolic enzymes appear to also moonlight as proteins involved in RNA processing.[99] These results suggest a possible novel function for these proteins in plant RNA metabolism as well.

Although evidence suggests RBP-A, -D and -I may play a role in storage protein gene expression, their specific functions remain unknown. The colocalization of RBP-A with microtubules and cortical cis-ER and RBP-D with actin filaments as well as prolamine and glutelin RNA binding activity for both could imply involvement in RNA transport, anchoring, or translational regulation. Their roles may vary, however, as evidenced by their temporal expression during seed development. RBP-A expression coincides with that of prolamine and glutelin protein accumulation, while RBP-D is more prevalent at the early stages of storage protein expression and RBP-I is expressed throughout seed development.[82] Current and future studies include identifying RNA targets of these RBPs using both biochemical techniques and motif prediction software, which may provide information on their general role in gene expression. The generation of RNAi lines as well as identification of genetic mutants will also prove useful. Furthermore, characterization of some of the other RBPs identified from the prolamine zipcode capture experiment[82] is underway, and preliminary evidence indicates their possible association with an RNP complex(es) containing RBP-A, -D and -I (Yang and Okita, unpublished results). Indeed, the existence of a complex of RBPs that associates with prolamine mRNAs may explain why little RNA binding specificity is exhibited in vitro. For example, although native RBP-A from a seed extract interacts with biotinylated prolamine cis-localization sequences under highly stringent binding conditions, the recombinant protein binds equally well to both zipcode and a control RNA (Crofts et al, unpublished results). This suggests other binding partners or factors are necessary for specific recognition, an observation reported for other eukaryotic RBPs.[29,100]

The evidence that some prolamine RNA binding proteins such as RBP-A, -D and *Os*TudorSN also bind glutelin RNA is of great interest. As previously mentioned, evidence suggests the prolamine and glutelin RNA localization pathways are interrelated and transport from the nucleus to the cortical ER may occur within a single RNP transport particle that splits en route to the PB-ER or cis-ER (Fig. 2). This suggests the existence of both common and specific trans-acting factors required for proper localization. Cytoplasmic, cytoskeleton-associated factors regulating glutelin gene expression are unknown, although efforts are ongoing to identify proteins that interact with the recently identified glutelin RNA cis-localization sequences.[52]

Analysis of RNA Mislocalization Mutants

As discussed previously, studies of rice genetic mutants indicate that the three RNA localization pathways (the two targeted pathways to the PB-ER and cis-ER and the default pathway to the cis-ER) are interrelated.[24,53] In the *glup2* mutant, prolamine mRNA is mislocalized to the cisternal ER, while in *glup4* and *glup6*, glutelin mRNA is partially mislocalized to the PB-ER. Interestingly, another mutant, *glup5*, exhibits mistargeting of globulin RNA to the cis-ER, yet prolamine mRNA transport to the PB-ER remains unaffected. This raises the possibility that there may be distinct mechanisms for localization to the PB-ER. Characterization of two of these RNA mislocalization mutants has provided important insight regarding the genes required for proper RNA transport and their potential roles. Incorrect RNA sorting in the non-allelic *glup4* and *glup6* mutants indicate there are at least two factors required for glutelin mRNA localization to the cis-ER. Map-based cloning efforts resulted in the identification of the *Glup4* gene as encoding for the small GTPase Rab5.[101] Most notable for its role in endosome formation in early endocytosis, Rab5 has also been implicated in endosome fusion, cargo recruitment, and movement of endosomes and organelles along the cytoskeleton.[102,103] In addition, Rab5 is an important factor in ER structuring.[104]

Further analysis of the *glup4* and *glup6* mutants revealed both glutelin RNA and protein are mistargeted in endosperm cells. Instead of being sorted solely to protein storage vacuoles, glutelin polypeptides are secreted as well as found within smaller protein bodies and large dilated membranous complexes (Fukuda et al, unpublished results). In *glup4*, this novel structure also contained protein markers for ER, Golgi, prevacuolar compartment, and plasma membrane, indicating a disruption in the general endomembrane architecture of the cell. A recent study of a rice mutant expressing truncated Rab5 found a similar effect on membrane organization within endosperm cells as well as mislocalization of protein storage vacuole markers, including glutelin, to small vesicles and multivesicular bodies.[105] In other plant systems, dominant negative mutants of Rab5 homologues suggest a role in vacuolar protein trafficking, as marker proteins are instead secreted or mislocalized within the cell.[106,107] These results suggest Rab5 plays an important role in protein sorting and maintenance of the endomembrane system in plants.

But what of its possible role in RNA localization? Partial mislocalization of glutelin mRNA to the PB-ER instead of the cis-ER in the *glup4* mutant indicates Rab5 may also be involved in RNA transport. This is supported by evidence that suggests membranes and their associated proteins do indeed play a role in RNA localization. In *Drosophila*, the ESCRT-II complex, initially found to be involved in endosomal sorting and formation of multivesicular bodies, is required for the targeted transport of bicoid mRNA to the anterior of the oocyte.[108] One component, VPS36, interacts with the bicoid 3'UTR zipcode in a sequence specific manner and is necessary for the recruitment of Staufen, a known RBP involved in bicoid RNA localization. Also in *Drosophila*, a genetic screen identified proteins involved in membrane fusion events, including homotypic fusion of the ER, as being necessary for oskar mRNA localization to the posterior pole of oocytes.[109] Membranes themselves, and in particular ER, have been identified as components of RNA transport as well. The association of gurken mRNA with nuclear membranes in *Drosophila* oocytes and dendritic mRNAs with ER in neurons is required for proper RNA localization.[110,111] In *Xenopus* oocytes, targeted transport of Vg1 mRNA to the vegetal cortex requires an interaction with ER membranes.[112] Furthermore, ER, Vg1 mRNA, and Vera, an RBP necessary for Vg1 localization, were all found to cofractionate, suggesting both ER and Vg1 mRNA are cotransported.[112] Similarly in yeast, transport of asymmetrically localized mRNAs to the bud likely occurs with cortical ER and associated RBPs in a cytoskeleton-dependent manner.[113,114] Whether rice seed storage protein mRNAs are similarly transported with ER membranes en route to the cortical region of the cell is unknown, but given that glutelin mRNA is mistargeted in the *glup4* and *glup6* mutants (ref. 55, Washida et al, unpublished results) and Rab5 is suggested to be involved in ER structuring,[104] an interesting possibility is raised regarding a role for Rab5 in RNA/ER colocalization. Perhaps glutelin mRNA transport occurs as part of RNP complexes associated with ER vesicles which fuse to the cisternal ER upon reaching the cortical region of the cell, events mediated by Rab5 (Fig. 3). The observation that prolamine mRNA localization to the PB-ER is not disrupted in *glup4* and *glup6* may indicate the activity of another Rab protein is required for prolamine mRNA transport.

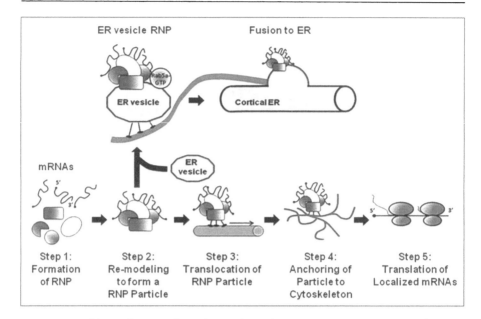

Figure 3. Possible mechanism of membrane-dependent storage protein mRNA localization. The model pathway of RNA localization proposed by Wilhelm and Vale[17] entails nuclear formation of the RNP complex, export to the cytoplasm, assembly of the RNP transport particle, cytoskeleton-dependent transport of the particle to its final destination, and translation. In developing rice seed, however, storage protein mRNA localization may require association of the RNP transport particle with ER vesicles that move along the cytoskeleton and ultimately fuse to the cortical ER prior to translation. These events may be mediated by GTP-activated Rab5 and other accessory proteins. Adapted from Wilhelm and Vale.[17]

Interestingly, the participation of a Rab protein in RNA transport has been previously reported. In *Drosophila* oocytes, oskar mRNA transport to the posterior pole is impaired in a *rab11* loss of function mutant and instead forms aggregates near the pole.[115,116] It has been proposed that Rab11, normally involved in vesicle trafficking to the plasma membrane, may transport oskar mRNA to its final destination along actin filaments, perhaps by participating in mRNA anchoring to vesicles or the cytoskeleton.[110] The localization of ER membranes and oskar mRNA to abnormal particulate structures in membrane fusion mutants support the idea that RNAs may be transported as part of a membrane complex.[109] Rab11 could also be involved in linking mRNA to the cytoskeleton, however, as it and other Rab proteins have been found to interact with motor proteins, either directly or indirectly, through effector proteins.[117] Effector proteins, which bind activated GTP-bound Rab, are responsible for the various Rab-dependent processes within the cell, including vesicle formation, transport, and fusion.[118] It is possible an effector protein is also specifically responsible for the RNA localization activity of Rab5 in rice and efforts are currently underway in our laboratory to identify potential candidates.

To better understand the effect of Rab5 on global gene expression events within plants, transcript and protein abundance of wildtype and *glup4* mutant seeds were compared using microarray and two-dimensional difference gel electrophoresis (2D-DIGE). Consistent with the altered membrane architecture and protein mis-sorting seen in *glup4* endosperm cells (Fukuda et al, unpublished results), a number of genes encoding cell wall biosynthetic enzymes, membrane-associated proteins and protein synthesis and modification proteins were differentially expressed in the mutant.[55] Interestingly, the expression of certain carbon metabolic enzymes, including those involved in starch biosynthesis, was also altered. The coordinated synthesis of seed storage reserves has long been observed, i.e., an increase in either storage protein or starch levels results in greater synthesis of the other.[119-122] Our results suggest that Rab5 may not only have a role in storage protein production, but also in starch biosynthesis as well. As Rab5 functions in both glutelin RNA and protein sorting (refs. 55, 106; Fukuda et al, unpublished results), it is possible that a similar Rab5 activity participates in starch biosynthetic gene expression, supporting the idea of co-regulated synthesis of storage reserves.

One protein of particular interest identified in the 2D-DIGE study that was downregulated in the *glup4* mutant was glyceraldehyde 3-phosphate dehydrogenase (GAPDH). A well-studied glycolytic enzyme, GAPDH is now widely accepted as a multifunctional protein with roles in DNA/RNA metabolism, cytoskeleton assembly, protein modification and membrane fusion, activities that are likely regulated by specific post-translational modifications.[123] In mammalian cells, GAPDH was found to interact with Rab2 and be necessary for vesicular protein trafficking between ER and Golgi.[124] A more recent study determined a bacterial GAPDH bound mammalian Rab5 and inhibited its GTP/GDP exchange activity as part of host cell infection.[125] As previously mentioned, co-immunoprecipitation experiments performed in our laboratory indicate GAPDH may be part of a multi-protein complex with RBPs identified as rice prolamine mRNA binding proteins (RBP-A, -I). Could rice GAPDH also be involved in gene expression in plants, perhaps as part of membrane-associated RNA transport complexes involving Rab proteins? The RNA-binding activity of GAPDH, which at least in some cases is redox regulated,[126,127] combined with its involvement in vesicle trafficking presents an intriguing scenario that requires further study.

Conclusion

Although the events surrounding intracellular RNA transport in plants are largely unknown, progress is being made. Continuing characterization of RBPs that interact with rice storage protein RNA cis-localization sequences will provide additional clues as to how this process is accomplished. This includes identifying interacting protein and RNA partners which may make up RNP transport particles. Elucidation of the effector protein responsible for Rab5's glutelin RNA localization activity as well as the factors causing RNA mislocalization in the *glup2* and *glup5* mutants will also aid in our understanding of the transport mechanism. Ultimately, these results will demonstrate whether components of the proposed storage protein RNA localization pathways are shared, if transport occurs within the same RNP transport particle, and which factors confer specificity in targeting to the PB-ER or cis-ER. Determination of RNA ligands for RBPs involved in localization will also shed light on whether functionally related sets of mRNAs are cotransported to similar subcellular regions as part of spatial and temporal regulation of protein expression and possible complex formation as in other eukaryotes.[25] The use of high throughput sequencing techniques to assess RNA ligands may also provide information as to whether noncoding RNAs are involved in this process.

Some of the more intriguing questions that remain are how prevalent active RNA transport is within plant cells and how it affects development. In rice endosperm alone, it is possible other localization pathways exist, such as for RNAs encoding carbon metabolic enzymes to ER near amyloplasts, which would allow for efficient protein targeting for starch biosynthesis. It is interesting to note the affect of the *glup4* mutation on the expression of certain starch biosynthetic enzymes, which may indicate RNA localization plays an important role in coordinating seed storage reserve production. Indeed, evidence for the inter-relatedness of starch and storage protein biosynthesis has been reported.[119-122] One possible model for this requiring further study is that these RNAs are (co) transported by a similar mechanism (i.e., shared components) to common sites on the cortical ER, where evidence from our laboratory suggests the majority of protein synthesis occurs in endosperm. Of course, this raises other exciting questions as to whether the differentiation of ER into uniquely functioning subdomains is a common form of regulation in protein synthesis.

What of the occurrence of RNA localization in other tissues or organisms, though? Similar to metazoans and yeast, it is likely targeted RNA transport is a major factor in the regulation of gene expression in processes such as embryogenesis, signaling, stress response and development. Although a few examples of RNA localization have been reported, one of the major hurdles to studying this phenomenon in plants has been the lack of high resolution in situ visualization techniques, which in other eukaryotic systems has led to incredible progress in the field. The development of new ways to monitor the movement of RNAs and RBPs within the cell as well as the discovery of new models for studying RNA localization in plants would be greatly beneficial. Continued focus on the functional elucidation of plant RBPs in general, however, will also be important for our overall understanding of gene expression events within the cell.

Acknowledgments

The authors thank John Wyrick for helpful discussion on the manuscript. This research was generously supported by National Science Foundation Grants IOB-0544469 and DBI-0605016; USDA grant 2006-35301-17043; and by a grant-in-aid for Scientific Research from the Japan Society for the Promotion of Science to (21380008).

References

1. Wilkinson MF, Shyu AB. Multifunctional regulatory proteins that control gene expression in both the nucleus and the cytoplasm. Bioessays 2001; 23(9):775-787.
2. Dreyfuss G, Kim VN, Kataoka N. Messenger-RNA-binding proteins and the messages they carry. Nat Rev Mol Cell Biol 2002; 3(3):195-205.
3. Lunde BM, Moore C, Varani G. RNA-binding proteins: modular design for efficient function. Nat Rev Mol Cell Biol 2007; 8(6):479-490.
4. Fedoroff NV. RNA-binding proteins in plants: the tip of an iceberg? Curr Opin Plant Biol 2002; 5(5):452-459.
5. Lorković ZJ, Barta A. Genome analysis: RNA recognition motif (RRM) and K homology (KH) domain RNA-binding proteins from the flowering plant Arabidopsis thaliana. Nucleic Acids Res 2002; 30(3):623-635.
6. Lorković ZJ. Role of plant RNA-binding proteins in development, stress response and genome organization. Trends Plant Sci 2009; 14(4):229-236.
7. Bailey-Serres J, Sorenson R, Juntawong P. Getting the message across: cytoplasmic ribonucleoprotein complexes. Trends Plant Sci 2009; 14(8):443-453.
8. St. Johnston D. Moving messages: the intracellular localization of mRNAs. Nat Rev Mol Cell Biol 2005; 6(5):363-375.
9. Martin KC, Ephrussi A. mRNA localization: gene expression in the spatial dimension. Cell 2009; 136(4):719-730.
10. Okita TW, Choi SB. mRNA localization in plants: targeting to the cell's cortical region and beyond. Curr Opin Plant Biol 2002; 5(6):553-559.
11. Jansen RP. mRNA localization: message on the move. Nat Rev Mol Cell Biol 2001; 2(4):247-256.
12. Du TG, Schmid M, Jansen RP. Why cells move messages: the biological functions of mRNA localization. Semin Cell Dev Biol 2007; 18(2):171-177.
13. Lécuyer E, Yoshida H, Parthasarathy N et al. Global analysis of mRNA localization reveals a prominent role in organizing cellular architecture and function. Cell 2007; 131(1):174-187.
14. Palacios IM. How does an mRNA find its way? Intracellular localisation of transcripts. Semin Cell Dev Biol 2007; 18(2):163-170.
15. Meignin C, Davis I. Transmitting the message: intracellular mRNA localization. Curr Opin Cell Biol 2010; 22(1):112-119.
16. Singer RH. RNA zipcodes for cytoplasmic addresses. Curr Biol 1993; 3(10):719-721.
17. Wilhelm JE, Vale RD. RNA on the move: the mRNA localization pathway. J Cell Biol 1993; 123(2):269-274.
18. Giorgi C, Moore MJ. The nuclear nurture and cytoplasmic nature of localized mRNPs. Semin Cell Dev Biol 2007; 18(2):186-193.
19. Lewis RA, Mowry KL. Ribonucleoprotein remodeling during RNA localization. Differentiation 2007; 75(6):507-518.
20. Krecic AM, Swanson MS. hnRNP complexes: composition, structure, and function. Curr Opin Cell Biol 1999; 11(3):363-371.
21. Besse F, Ephrussi A. Translational control of localized mRNAs: restricting protein synthesis in space and time. Nat Rev Mol Cell Biol 2008; 9(12):971-980.
22. Kloc M, Zearfoss NR, Etkin LD. Mechanisms of subcellular mRNA localization. Cell 2002; 108(4):533-544.
23. López de Heredia M, Jansen RP. mRNA localization and the cytoskeleton. Curr Opin Cell Biol 2004; 16(1):80-85.

24. Crofts AJ, Washida H, Okita TW et al. Targeting of proteins to endoplasmic reticulum-derived compartments in plants. The importance of RNA localization. Plant Physiol 2004; 136(3):3414-3419.
25. Lécuyer E, Yoshida H, Krause HM. Global implications of mRNA localization pathways in cellular organization. Curr Opin Cell Biol 2009; 21(3):409-415.
26. Jambhekar A, Derisi JL. Cis-acting determinants of asymmetric, cytoplasmic RNA transport. RNA 2007; 13(5):625-642.
27. Mowry KL. Complex formation between stage-specific oocyte factors and a Xenopus mRNA localization element. Proc Natl Acad Sci USA 1996; 93(25):14608-14613.
28. Zhao WM, Jiang C, Kroll TT et al. A proline-rich protein binds to the localization element of Xenopus Vg1 mRNA and to ligands involved in actin polymerization. EMBO J 2001; 20(9):2315-2325.
29. Arn EA, Cha BJ, Theurkauf WE et al. Recognition of a bicoid mRNA localization signal by a protein complex containing Swallow, Nod, and RNA binding proteins. Dev Cell 2003; 4(1):41-51.
30. Haywood V, Kragler F, Lucas WJ. Plasmodesmata: pathways for protein and ribonucleoprotein signaling. Plant Cell 2002; 14:S303-S325.
31. Kehr J, Buhtz A. Long distance transport and movement of RNA through the phloem. J Exp Bot 2008; 59(1):85-92.
32. Yoo BC, Kragler F, Varkonyi-Gasic E et al. A systemic small RNA signaling system in plants. Plant Cell 2004; 16(8):1979-2000.
33. Ham BK, Brandom JL, Xoconostle-Cázares B et al. A polypyrimidine tract binding protein, pumkin RBP50, forms the basis of a phloem-mobile ribonucleoprotein complex. Plant Cell 2009; 21(1):197-215.
34. Bouget FY, Gerttula S, Shaw SL et al. Localization of actin mRNA during the establishment of cell polarity and early cell divisions in Fucus embryos. Plant Cell 1996; 8(2):189-201.
35. Serikawa KA, Porterfield DM, Mandoli DF. Asymmetric subcellular mRNA distribution correlates with carbonic anhydrase activity in Acetabularia acetabulum. Plant Physiol 2001; 125(2):900-911.
36. Baluska F, Salaj J, Mathur J et al. Root hair formation: F-actin-dependent tip growth is initiated by local assembly of profilin-supported F-actin meshworks accumulated within expansin-enriched bulges. Dev Biol 2000; 227(2):618-632.
37. Im KH, Cosgrove DJ, Jones AM. Subcellular localization of expansin mRNA in xylem cells. Plant Physiol 2000; 123(2):463-470.
38. Nicolaï M, Duprat A, Sormani R et al. Higher plant chloroplasts import the mRNA coding for the eucaryotic translation initiation factor 4E. FEBS Lett 2007; 581(21):3921-3926.
39. Uniacke J, Zerges W. Chloroplast protein targeting involves localized translation in Chlamydomonas. Proc Natl Acad Sci USA 2009; 106(5):1439-1444.
40. Michaud M, Maréchal-Drouard L, Duchêne AM. RNA trafficking in plant cells: targeting of cytosolic mRNAs to the mitochondrial surface. Plant Mol Biol 2010; 73(6):697-704.
41. Samaj J, Salaj J, Obert B et al. Calreticulin mRNA and protein are localized to the protein bodies in storage maize callus cells. Plant Cell Rep 2008; 27(2):231-239.
42. Muench DG, Okita TW. The storage proteins of rice and oat. In: Larkins BA, Vasil IK, eds. Cellular and Molecular Biology of Plant Development. The Netherlands: Kluwer Academic Press, 1997:289-330.
43. Tanaka K, Sugimoto T, Ogawa M et al. Isolation and characterization of two types of protein bodies in the rice endosperm. Agric Biol Chem 1980; 44(7):1633-1639.
44. Krishnan HB, Franceschi VR, Okita TW. Immunochemical studies on the role of the Golgi complex in protein body formation in rice seeds. Planta 1986; 169(4):471-480.
45. Yamagata H, Tanaka K. The site of synthesis and accumulation of rice storage proteins. Plant Cell Physiol 1986; 27(1):135-145.
46. Muench DG, Chuong SD, Franceschi VR et al. Developing prolamine protein bodies are associated with the cortical cytoskeleton in rice endosperm cells. Planta 2000; 211(2):227-238.
47. Clore AM, Dannenhoffer JM, Larkins BA. EF-1α is associated with a cytoskeletal network surrounding protein bodies in maize endosperm cells. Plant Cell 1996; 8(11):2003-2014.
48. Li X, Franceschi VR, Okita TW. Segregation of storage protein mRNAs on the rough endoplasmic reticulum membranes of rice endosperm cells. Cell 1993; 72(6):869-879.
49. Choi SB, Wang C, Muench DG et al. Messenger RNA targeting of rice seed storage proteins to specific ER subdomains. Nature 2000; 407(6805):765-767.
50. Hamada S, Ishiyama K, Choi SB et al. The transport of prolamine RNAs to prolamine protein bodies in living rice endosperm cells. Plant Cell 2003; 15(10):2253-2264.
51. Hamada S, Ishiyama K, Sakulsingharoj C et al. Dual regulated RNA transport pathways to the cortical region in developing rice endosperm. Plant Cell 2003; 15(10):2265-2272.
52. Washida H, Kaneko S, Crofts N et al. Identification of cis-localization elements that target glutelin RNAs to a specific subdomain of the cortical endoplasmic reticulum in rice endosperm cells. Plant Cell Physiol 2009; 50(9):1710-1714.

53. Crofts AJ, Washida H, Okita TW et al. The role of mRNA and protein sorting in seed storage protein synthesis, transport, and deposition. Biochem Cell Biol 2005; 83(6):728-737.
54. Yamagata H, Sugimoto T, Tanaka K et al. Biosynthesis of storage proteins in developing rice seeds. Plant Physiol 1982; 70(4):1094-1100.
55. Doroshenk KA, Crofts AJ, Washida H et al. Characterization of the rice glup4 mutant suggests a role for the small GTPase Rab5 in the biosynthesis of carbon and nitrogen storage reserves in developing endosperm. Breeding Sci 2010; 60(5):556-567.
56. Larkins BA, Hurkman WJ. Synthesis and deposition of zein in protein bodies of maize endosperm. Plant Physiol 1978; 62(2):256-263.
57. Kim CS, Woo YM, Clore AM et al. Zein protein interactions, rather than the asymmetric distribution of zein mRNAs on the endoplasmic reticulum membranes, influence protein body formation in maize endosperm. Plant Cell 2002; 14(3):655-672.
58. Washida H, Sugino A, Messing J et al. Asymmetric localization of seed storage protein RNAs to distinct subdomains of the endoplasmic reticulum in developing maize endosperm cells. Plant Cell Physiol 2004; 45(12):1830-1837.
59. Washida H, Sugino A, Kaneko S et al. Identification of cis-localization elements of the maize 10-kDa delta-zein and their use in targeting RNAs to specific cortical endoplasmic reticulum subdomains. Plant J 2009; 60(1):146-155.
60. Chuong SD, Good AG, Taylor GJ et al. Large-scale identification of tubulin-binding proteins provides insight on subcellular trafficking, metabolic channeling, and signaling in plant cells. Mol Cell Proteomics 2004; 3(10):970-983.
61. Chuong SD, Mullen RT, Muench DG. Identification of a rice RNA- and microtubule-binding protein as the multifunctional protein, a peroxisomal enzyme involved in the beta-oxidation of fatty acids. J Biol Chem 2002; 277(4):2419-2429.
62. Chuong SD, Mullen RT, Muench DG. The peroxisomal multifunctional protein interacts with cortical microtubules in plant cells. BMC Cell Biol 2005; 6:40.
63. Tam PP, Barrette-Ng IH, Simon DM et al. The Puf family of RNA-binding proteins in plants: phylogeny, structural modeling, activity and subcellular localization. BMC Plant Biol 2010; 10:44-63.
64. Ma L, Xie B, Hong Z et al. A novel RNA-binding protein associated with cell plate formation. Plant Physiol 2008; 148(1):223-234.
65. Guo F, Yu L, Watkins S et al. Orientation of microtubules suggests a role in mRNA transportation in fertilized eggs of Chinese pine (Pinus tabulaeformis). Protoplasma 2007; 231(3-4):239-243.
66. Heyl A, Muth J, Santandrea G et al. A transcript encoding a nucleic-acid binding protein specifically expressed in maize seeds. Mol Genet Genomics 2001; 266(2):180-189.
67. Muench DG, Wu Y, Coughlan SJ et al. Evidence for a cytoskeleton-associated binding site involved in prolamine mRNA localization to the protein bodies in rice endosperm. Plant Physiol 1998; 116(2):559-569.
68. Wu Y, Muench DG, Kim YT et al. Identification of polypeptides associated with an enriched cytoskeleton-protein body fraction from developing rice endosperm. Plant Cell Physiol 1998; 39(12):1251-1257.
69. Sami-Subbu R, Muench DG, Okita TW. A cytoskeleton-associated RNA-binding protein binds to the untranslated regions of prolamine mRNA and to poly(A). Plant Sci 2000; 152(2):115-122.
70. Sami-Subbu R, Choi SB, Wu Y et al. Identification of a cytoskeleton-associated 120 kDa RNA-binding protein in developing rice seeds. Plant Mol Biol 2001; 46(1):79-88.
71. Tong X, Drapkin R, Yalamanchili R et al. The Epstein-Barr virus nuclear protein 2 acidic domain forms a complex with a novel cellular coactivator that can interact with TFIIE. Mol Cell Biol 1995; 15(9):4735-4744.
72. Leverson JD, Koskinen PJ, Orrico FC et al. Pim-1 kinase and p100 cooperate to enhance c-Myb activity. Mol Cell 1998; 2(4):417-425.
73. Yang J, Aittomäki S, Pesu M et al. Identification of p100 as a coactivator for STAT6 that bridges STAT6 with RNA polymerase II. EMBO J 2002; 21(18):4950-4958.
74. Caudy AA, Ketting RF, Hammond SM et al. A micrococcal nuclease homologue in RNAi effector complexes. Nature 2003; 425(6956):411-414.
75. Scadden AD. The RISC subunit Tudor-SN binds to hyper-edited double-stranded RNA and promotes its cleavage. Nat Struct Mol Biol 2005; 12(6):489-496.
76. Liu S, Jia J, Gao Y et al. The AtTudor2, a protein with SN-Tudor domains, is involved in control of seed germination in Arabidopsis. Planta 2010; 232(1):197-207.
77. Dit Frey NF, Muller P, Jammes F et al. The RNA binding protein Tudor-SN is essential for stress tolerance and stabilizes levels of stress-responsive mRNAs encoding secreted proteins in Arabidopsis. Plant Cell 2010; 22(5):1575-1591.
78. Wang C, Washida H, Crofts AJ et al. The cytoplasmic-localized, cytoskeletal-associated RNA binding protein OsTudor-SN: evidence for an essential role in storage protein RNA transport and localization. Plant J 2008; 55(3):443-454.

79. Doroshenk KA, Crofts AJ, Morris RT et al. Proteomic analysis of cytoskeleton-associated RNA binding proteins in developing rice seed. J Proteome Res 2009; 8(10):4641-4653.
80. Cote CA, Gautreau D, Denegre JM et al. A Xenopus protein related to hnRNP I has a role in cytoplasmic RNA localization. Mol Cell 1999; 4(3):431-437.
81. Besse F, López de Quinto S, Marchand V et al. Drosophila PTB promotes formation of high-order RNP particles and represses oskar translation. Genes Dev 2009; 23(2):195-207.
82. Crofts AJ, Crofts N, Whitelegge JP et al. Isolation and identification of cytoskeleton-associated prolamine mRNA binding proteins from developing rice seeds. Planta 2010; 231(6):1261-1276.
83. Chaikam V, Karlson DT. Comparison of structure, function, and regulation of plant cold shock domain proteins to bacterial and animal cold shock domain proteins. BMB Rep 2010; 43(1):1-8.
84. Wilhelm JE, Mansfield J, Hom-Booher N et al. Isolation of a ribonucleoprotein complex involved in mRNA localization in Drosophila oocytes. J Cell Biol 2000; 148(3):427-440.
85. Lorković ZJ, Wieczorek Kirk DA, Klahre U et al. RBP45 and RBP47, two oligouridylate-specific hnRNP-like proteins interacting with poly(A)+ RNA in nuclei of plant cells. RNA 2000; 6(11):1610-1624.
86. Weber C, Nover L, Fauth M. Plant stress granules and mRNA processing bodies are distinct from heat stress granules. Plant J 2008; 56(4):517-530.
87. Morris RT, Doroshenk KA, Crofts AJ et al. RiceRBP: a database of experimentally identified RNA-binding proteins in Oryza sativa L. Plant Sci 2011; 180(2)204-211.
88. Walker NS, Stiffler N, Barkan A. POGs/PlantRBP: a resource for comparative genomics in plants. Nucleic Acid Res 2007; 35:D852-D856.
89. Norvell A, Kelley RL, Wehr K et al. Specific isoforms of squid, a Drosophila hnRNP, perform distinct roles in Gurken localization during oogenesis. Genes Dev 1999; 13(7):864-876.
90. Huynh JR, Munro TP, Smith-Litière K et al. The Drosophila hnRNPA/B homolog, Hrp48, is specifically required for a distinct step in osk mRNA localization. Dev Cell 2004; 6(5):625-635.
91. Yano T, López de Quinto S, Matsui Y et al. Hrp48, a Drosophila hnRNPA/B homolog, binds and regulates translation of oskar mRNA. Dev Cell 2004; 6(5):637-648.
92. Delanoue R, Herpers B, Soetaert J et al. Drosophila squid/hnRNP helps dynein switch from a gurken mRNA transport motor to an unltrastructural static anchor in sponge bodies. Dev Cell 2007; 13(4):523-538.
93. Norvell A, Debec A, Finch D et al. Squid is required for efficient posterior localization of oskar mRNA during Drosophila oogenesis. Dev Genes Evol 2005; 215(7):340-349.
94. Matunis EL, Kelley R, Dreyfuss G. Essential role for a heterogeneous nuclear ribonucleoprotein (hnRNP) in oogenesis: hrp40 is absent from the germ line in the dorsoventral mutant squid. Proc Natl Acad Sci 1999; 91(7):2781-2784.
95. Siebel CW, Kanaar R, Rio DC. Regulation of tissue-specific P-element pre-mRNA splicing requires the RNA-binding protein PSI. Genes Dev 1994; 8(14):1713-1725.
96. Hammond LE, Rudner DZ, Kanaar R et al. Mutations in the hrp48 gene, which encodes a Drosophila heterogeneous nuclear ribonucleoprotein particle protein, cause lethality and developmental defects and affect P-element third-intron splicing in vivo. Mol Cell Biol 1997; 17(12):7260-7267.
97. Goodrich JS, Clouse KN, Schupbach T. Hrb27C, Sqd and Otu cooperatively regulate gurken RNA localization and mediate nurse cell chromosome dispersion in Drosophila oogenesis. Development 2004; 131(9):1949-1958.
98. Nagy E, Henics T, Eckert M et al. Identification of the NAD⁺-binding fold of glyceraldehyde-3-phosphate dehydrogenase as a novel RNA-binding domain. Biochem Biophys Res Commun 2000; 275(2):252-260.
99. Cieśla J. Metabolic enzymes that bind RNA: yet another level of cellular regulatory network? Acta Biochimica Polonica 2006; 53(1):11-32.
100. Singh R, Valcárcel J. Building specificity with nonspecific RNA-binding proteins. Nat Struct Mol Biol 2005; 12(8):645-653.
101. Satoh-Cruz M, Fukuda M, Ogawa M et al. Glup4 gene encodes small GTPase Rab5a in rice. Rice Genet Newslet 2010; 25:48-49.
102. Zerial M, McBride H. Rab proteins as membrane organizers. Nat Rev Mol Cell Biol 2001; 2(2):107-117.
103. Van der Bliek AM. A sixth sense for Rab 5. Nat Cell Biol 2005; 7(6):548-550.
104. Audhya A, Desai A, Oegema K. A role for Rab5 in structuring the endoplasmic reticulum. J Cell Biol 2007; 178(1):43-56.
105. Wang Y, Ren Y, Liu X et al. OsRab5a regulates endomembrane organization and storage protein trafficking in rice endosperm cells. Plant J 2010; 64(5)812-824.
106. Sohn EJ, Kim ES, Zhao M et al. Rha1, an Arabidopsis Rab5 homolog, plays a critical role in the vacuolar trafficking of soluble cargo proteins. Plant Cell 2003; 15(5):1057-1070.
107. Kotzer AM, Brandizzi F, Neumann U et al. AtRabF2b (Ara7) acts on the vacuolar trafficking pathway in tobacco leaf epidermal cells. J Cell Sci 2004; 117(26):6377-6389.

108. Irion U, St. Johnston D. bicoid RNA localization requires specific binding of an endosomal sorting complex. Nature 2007; 445(7127):554-558.
109. Ruden DM, Sollars V, Wang X et al. Membrane fusion proteins are required for oskar mRNA localization in the Drosophila egg chamber. Dev Biol 2000; 218(2):314-325.
110. Cohen RS. The role of membranes and membrane trafficking in RNA localization. Biol Cell 2005; 97(1):5-18.
111. Gerst JE. Message on the web: mRNA and ER co-trafficking. Trends Cell Biol 2008; 18(2):68-76.
112. Deshler JO, Highett MI, Schnapp BJ. Localization of Xenopus Vg1 mRNA by Vera protein and the endoplasmic reticulum. Science 1997; 276(5315):1128-1131.
113. Schmid M, Jaedicke A, Du TG et al. Coordination of endoplasmic reticulum and mRNA localization to the yeast bud. Curr Biol 2006; 16(15):1538-1543.
114. Aronov S, Gelin-Licht R, Zipor G et al. mRNAs encoding polarity and exocytosis factors are cotransported with the cortical endoplasmic reticulum to the incipient bud in Saccharomyces cerevisiae. Mol Cell Biol 2007; 27(9):3441-3455.
115. Dollar G, Struckhoff E, Michaud J et al. Rab11 polarization of the Drosophila oocyte: a novel link between membrane trafficking, microtubule organization, and oskar mRNA localization and translation. Development 2002; 129(2):517-526.
116. Jankovics F, Sinka R, Erdélyi M. An interaction type of genetic screen reveals a role of the Rab11 gene in oskar mRNA localization in the developing Drosophila melanogaster oocyte. Genetics 2001; 158(3):1177-1188.
117. Jordens I, Marsman M, Kuijl C et al. Rab proteins, connecting transport and vesicle fusion. Traffic 2005; 6(12):1070-1077.
118. Grosshans BL, Ortiz D, Novick P. Rabs and their effectors: achieving specificity in membrane traffic. Proc Natl Acad Sci USA 2006; 103(32):11821-11827.
119. Nelson OE. Genetic control of polysaccharide and storage protein synthesis in endosperms of barley, maize and sorghum. In: Pomeranz Y, ed. Advances in Cereal Science and Technology. St. Paul: American Association of Cereal Chemists, 1982:41-71.
120. Stark DM, Timmerman KP, Barry GF et al. Regulation of the amount of starch in plant tissues by ADP glucose pyrophosphorylase. Science 1992; 258(5080):287-292.
121. Muller-Rober B, Sonnewald U, Willmitzer L. Inhibition of the ADP-glucose pyrophosphorylase in transgenic potatoes leads to sugar-storing tubers and influences tuber formation and expression of tuber storage protein genes. EMBO J 1992; 11(4):1229-1238.
122. Giroux MJ, Shaw J, Barry G et al. A single gene mutation that increases maize seed weight. Proc Natl Acad Sci USA 1996; 93(12):5824-5829.
123. Sirover MA. New insights into an old protein: the functional diversity of mammalian glyceraldehyde-3-phosphate dehydrogenase. Biochim Biophys Acta 1999; 1432(2):159-184.
124. Tisdale EJ, Kelly C, Artalego CR. Glyceraldehyde-3-phosphate dehydrogenase interacts with Rab2 and plays an essential role in endoplasmic reticulum to Golgi transport exclusive of its glycolytic activity. J Biol Chem 2004; 279(52):54046-54052.
125. Alvarez-Dominguez C, Madrazo-Toca F, Fernandez-Prieto L et al. Characterization of a Listeria monocytogenes protein interfering with Rab5a. Traffic 2008; 9(3):325-337.
126. Hwang NR, Yim SH, Kim YM et al. Oxidative modifications of glyceraldehyde-3-phosphate dehydrogenase play a key role in its multiple cellular functions. Biochem J 2009; 423(2):253-264.
127. Rodríguez-Pascual F, Redondo-Horcajo M, Magán-Marchal N et al. Glyceraldehyde-3-phosphate dehydrogenase regulates endothelin-1 expression by a novel, redox-sensitive mechanism involving mRNA stability. Mol Cell Biol 2008; 28(23):7139-7155.

CHAPTER 7

RNA Binding Domains and RNA Recognition by RNA-Editing Machineries

Michael F. Jantsch and Cornelia Vesely*

Abstract

RNA editing was originally believed to be a rare event occurring mostly in organelles of peculiar organisms. Advances in bioinformatics, the vast number of transcript and genomic sequences, together with novel sequencing technologies, however, has led to a dramatic change of this view. Today, RNA-editing has been reported in every kingdom of life and has been shown to be very abundant process, significantly contributing to transcriptome diversity in many cases. Mechanisms of RNA editing vary considerably, depending on the organismic group. In this chapter we give an overview on the machineries and mechanisms involved in RNA editing and substrate recognition.

Introduction

Beadle and Tatums postulate of one gene encoding one polypeptide has been a central part of genetics classroom teaching. Based on this postulate it was assumed that one gene would be transcribed into one RNA which would then be translated into one protein.[1] It was not until the late 1970s that this dogma was first challenged by the discovery of alternative splicing, first in viral transcripts, later in mammalian genes and subsequently in almost all eukaryotes.[2-4]

The second challenge to the one gene-one protein hypothesis came from the discovery of RNA-editing. Here, a plethora of mechanisms lead to the post transcriptional alteration of genetic information at the RNA level.

Initially discovered in mitochondrial genomes as a base insertion/deletion mechanism RNA editing was later also found in nuclear transcripts where base deamination is the most prominent mechanism.[5-7]

What is common to both, alternative splicing and RNA-editing is its role in transcriptome diversification: Sequencing of several genomes has revealed that organismic and genomic complexity do not correlate. Instead, a surprisingly similar number of genes can be found in organisms as diverse as *C. elegans* and humans.[8] Nonetheless, while the increase in gene number is only modest throughout evolution, proteomic complexity varies dramatically and seems to be largest in complex organs such as the mammalian brain.[9]

*Department of Chromosome Biology, Max F. Perutz Laboratories, University of Vienna, Vienna, Austria.
Corresponding Author: Michael F. Jantsch—Email: michael.jantsch@univie.ac.at

RNA Binding Proteins, edited by Zdravko J. Lorković.

Base Insertion/Deletion

Nucleotide insertion and/or deletion editing all affect mitochondrial RNA of lower eukaryotes. Insertion/deletion RNA editing can be a posttranscriptional modification of the RNA sequence, like in kinetoplastid RNA editing. Alternatively, insertion/deletion editing can occur cotranscriptionally, like in myxomycetes.

Kinetoplastid RNA Editing

Trypanosomes are unicellular eukaryotic organisms that belong to the kinetoplastid protozoa. Due to the fact that many of them are pathogenic parasites considerable research efforts focused on these organisms. The special feature of the kinetoplastid protozoa is the kinetoplast, a disk shaped granule of concatenated mitochondrial DNA circles located at the base of the flagellum. Most of the mitochondrial mRNAs get extensively edited by the insertion of—or less frequently—the deletion of uridines. Since the discovery of kinetoplastid editing in 1986 by Benne and coworkers the complex machinery that performs U insertion and deletions has been extensively characterized. The current model for kRNA editing includes different multiprotein complexes that act in the intricate cleavage/ligation mechanism.

The posttranscriptional insertion and deletion of uridine nucleotides in the pre-edited mRNAs (encoded in the kinetoplastid DNA maxicircles) is templated by small transacting ~60 nucleotides long guide RNAs (gRNAs) primarily encoded in the mitochondrial minicircles.[10,11] Guide RNAs basepair with short segments of the pre-mRNAs, whereby unpaired A (or G) residues in gRNAs provide the information for the number and location of uridines to be inserted or deleted.[12] gRNAs have oligo-U tails at their 3′ end that are added posttranscriptionally by the terminal uridyl transferase KRET1 and possibly serve as a stabilizer for the gRNA-mRNA interaction.[13,14] Moreover, gRNAs have a conserved secondary structure at their 5′ end, called the anchor sequence. This sequence contains two stem-loop regions that form an anchor duplex with the mRNA to initiate editing.[15] Thus, the direction of editing is always from the 3′ to 5′ end of the mRNA.

A complex that sediments at ~20 Svedberg (20S), and correspondingly called the '20S editosome', contains the four required key enzyme activities and is able to catalyze in vitro editing. The editing process is composed of three subsequent steps: Endonucleolytic cleavage of the pre-mRNA, uridine addition by TUTase or removal by uridine-specific exonuclease and finally ligation to rejoin the two halves of the pre-mRNA. Recently, evidence for the presence of a 3′ nucleotidyl phosphatase activity in the uridine deletion editing was found (Fig. 1A).[16]

Evidence exists for the presence of three different types of 20S editosomes that share a common core of 12 proteins (RNA editing core complex-RECC, or ligase complex).[17] The common set of proteins includes four proteins containing oligonucleotide/oligosaccharide binding (OB)-fold motifs (KREPA3-6-kinetoplastid RNA editing proteins [KREPs]) and two proteins with U1-like zinc finger motifs (KREPB4 and KREPB5).[18] Additionally, the core also comprises two subcomplexes, one for insertion and one for deletion editing: In the insertion complex the TUTase (KRET2) is linked by the (OB)-fold containing protein KREPA1 to the ligase (KREL2), while in the deletion complex the 3′ exonuclease KREX2 is linked to the ligase by KREPA2.[19] Proteins that differ between the three editosomes are part of a so called endonuclease subcomplex, which is built by the RNase III type endonucleases KREN1-3 and KREPB6-8.[17] One of the three editosomes contains an additional U-specific exonuclease (KREX1).[20] Within the editosomes, the KREPA and KREPB proteins form a network connecting the insertion with the deletion subcomplex via the OB-fold motifs of the KREPA proteins.[19]

In addition to the 20S editosome other factors indirectly regulate the efficiency of editing and editosome stability. One of these is the KRET1 containing TUT-II complex which performs the already mentioned addition of oligo-U tails to the 3′ ends of gRNAs.[14] An additional factor is the MRP1/MRP2 complex which plays a role in the association of the gRNA with its corresponding mRNA. MRP1 and MRP2 proteins form a heterotetrameric structure with a positively charged patch to which the gRNAs anneal.[21,22]

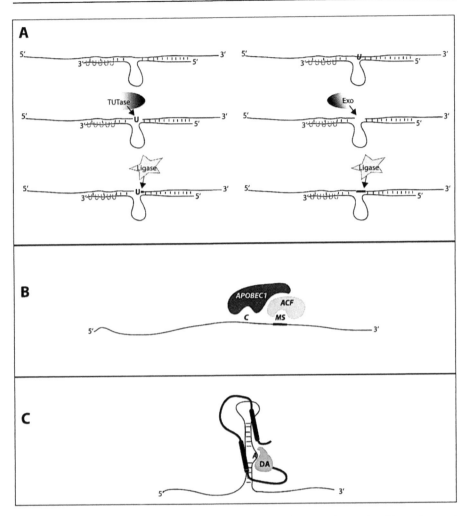

Figure 1. Different mechanisms of RNA recognition and RNA-editing. A) U-insertion and U-deletion editing as found in kinetoplastids is guided by short guide RNAs containing a U tail (lower strand) that basepair with the substrate RNA (upper strand). In insertion editing (left side) a cut is introduced 5' to the first nucleotide that basepairs with the 5' end of the guide RNA. A terminal uridylyl transferase (TUTase). The number of U residues added is determined by the ability of the newly added nucleotides to basepair with the guide RNA. Finally, the remaining gap is sealed by a ligase reaction. During U removal (right side) after cleavage of the substrate RNA a mismatched U residue is removed from the substrate RNA via an exonuclease (Exo). The remaining gap is again closed in a ligase reaction. B) Cytidine deamination by APOBEC1. In apolipoprotein B mRNA a mooring sequence (MS) is recognizec by the multi-protein apobec complementation factor (ACF). ACF interacts with ABOBEC1 guiding it to the C to be deaminated. Many recently identified APOBEC1 editing sites have a conserved region flanking the editing site that may be important to target the enzyme to its substrate nucleotide. C) Adenosine deamination by ADARs. A double-stranded structure is recognized by two or more double stranded RNA binding domains (grey cylinders). The dsRBDs help to position the deaminase domain (DA) close to the adenosine to be deaminated which is frequently not involved in basepairing.

Moreover, the mitochondrial RNA-binding protein RBP16 was shown to associate with gRNAs by binding the gRNA oligo-U-tail. It most probably acts as an accessory factor regulating editing and stability of mitochondrial mRNAs.[23] Recently, a mitochondrial RNA binding complex 1 (MRB1) was found that consists of 16 proteins, four of them have an RNA binding motif and one has an RNA helicase motif. MRB1 associates with the proteins TbRGG1 and TbRGG2. TbRGG1 contains an RGG RNA binding motif. The same is true for TbRGG2, which additionally harbors a C-terminal RRM. TbRGG1 is thought to function in stabilizing edited RNAs or editing efficiency.[24,25]

Myxomycetes

Besides trypanosomes, another group of protists namely myxomycetes (slime molds) show editing via base insertions or deletions in their mitochondria. While trypanosomes show post-transcriptional RNA editing, cotranscriptional editing is found in myxomyceta. Here, editing is characterized by an addition (or removal) of nontemplated nucleotides to the 3' end of the RNA during transcription. *Physarum polycephalum* can use up to three different types of editing within a single transcript: C-, U- and dinucleotide-insertions,[26] A-deletion as well as C to U conversion.[27,28] For example, in the cytochrome *c* oxidase subunit I transcript about 60 C-insertions, one U-insertion, three dinucleotide insertions and four C-to-U conversions take place.[28] So far, most mitochondrial editing helps to generate ORFs capable of directing protein synthesis.

Unlike in kinetoplastids no guide RNAs have been discovered in myxomyceta. To date, it is still not clear how the nontemplated addition (or the less frequent deletion) of nucleotides is accomplished in myxomycete mitochondria. Several pieces of evidence for the dependence of editing on the mitochondrial RNA polymerase (mtRNAP) itself have been found, although also other transacting factors are thought to be involved.[29] In vitro studies showed that the insertional type of editing happens cotranscriptionally. This cotranscriptional editing is however not due to stuttering of the RNA polymerase, like in viral cases of RNA editing,[30] but occurs through an addition of nontemplated nucleotides to the 3' end of the nascent RNA by the mitochondrial RNA polymerase.[31,32] Still, little is known about the attributes that define the location of editing or the identity of nucleotide(s) to be inserted. In 2009, the site of editing was more defined by Gott and coworkers: Bases up to nine nucleotides up and downstream of the editing site were shown to be both necessary and sufficient to allow insertional editing.[33]

However, not only mRNAs but also rRNAs and tRNAs are targets of editing, which results in changes in secondary and tertiary structures of mitochondrial tRNAs and rRNAs. Recently, also the insertion of single Gs into the 5'end of two mitochondrial tRNAs has been shown. Interestingly, this insertion is more likely to occur via a posttranscriptional mechanism that might be similar to the mechanism of *Acanthamoeba castellanii* and *Spizellomyces punctatus* tRNA editing (see below).[34]

5' tRNA Editing in Mitochondria of Protozoa

5' tRNA editing occurs in the mitochondria of the distantly related amoeboid protozoon *Acanthamoeba castellanii*[35] and the chytridiomycete fungus *Spizellomyces punctatus*,[36] as well as in the freshwater nanoplankton *Hyaloraphidium curvatum*.[37] tRNA editing reconstitutes canonical tRNA structures by exchanging mismatched nucleotides at the first three 5' positions for nucleotides that can form Watson-Crick base pairs with their counterparts. In *Acanthamoeba castellanii* at least two enzymes seem to be responsible for tRNA editing: An endonuclease and/or 5'-to-3' exonuclease for removal of mismatched nucleotides and a template-directed 3'-to-5' nucleotidyltransferase.[38] In vitro studies in *Spizellomyces punctatus* showed that tRNA editing uses the 3'-sequences of the acceptor stem as a template for the insertion of nucleotides in a 3' to 5' direction. Moreover, it was shown that editing functions independently of tRNA 3' processing.[39]

Dinoflagellates

Dinoflagellate mitochondrial and chloroplast RNAs also undergo substitutional editing. Other than in plant organelle editing and editing in animals, many different types of base conversions are observed. In fact only 3 of the possible 12 nucleotide changes have not been found so far.

Some editing events, especially A to G and C to U changes, occur in high frequency. For example, some chloroplast minicircle transcripts of *Ceratium horridum* together contain several hundred sites where seven different types of editing occur.[40] Editing affects protein coding genes as well as rRNA genes and occurs in clusters rather than evenly distributed over the genome. However, the mechanisms of editing in dinoflagellates remain unidentified. In coding transcripts editing occurs mostly at the first or second codon position favouring some particular amino acid transitions. Nonetheless, the biological role, especially of rRNA editing is still elusive.[41]

Base Modification

RNA editing by base conversion has mainly been reported from metazoa and plants. However, in recent years base conversion was also found to occur in dinoflagellates and placozoa. Different forms of modifications have been discovered so far: The edited base is modified by either a deamination (A to I and C to U) or more rarely an amination (U to C) reaction. Additionally, purine-to-pyrimidine and pyrimidine-to-purine changes have been found.

C to U Editing

APOBECs

APOBEC proteins form a zinc-dependent cytidine deaminase enzyme family that functions in deaminating single-stranded DNA or RNA. The only member of the APOBEC protein family that is known to act in RNA editing is APOBEC1, which was identified by its role in lipid metabolism. Through tissue-specifically regulated deamination of apolipoprotein B (apoB) mRNA at cytidine 6666 two isoforms of the apoB protein are generated as the deamination creates a $\underline{U}$AA stop codon from the genomically encoded $\underline{C}$AA codon for glutamine.[42] The full-length protein (apoB100) forms the lipoprotein component of low-density lipoprotein particles. The short isoform (apoB48) is essential for the formation of chylomicron lipoprotein particles and the transport of lipid in the small intestine. Moreover, APOBEC1 is also responsible for hyperediting of Nat1 and $\underline{C}$GA (Arg) to $\underline{U}$GA (Stop) editing in the neurofibromatosis-1 (NF1) mRNA.[43,44]

From the structure of the related APOBEC3G it is known that the catalytic center harbors an H[AV]E-x[24-36]-PCxxC motif (x = any amino acid) in which the H and the two Cs coordinate the zinc ion. The to be edited C is positioned in this pocket and is deaminated through nucleophilic attack.[45]

APOBEC1 forms an editing holoenzyme with APOBEC1 complementation factor (ACF) that plays the major role in editing site recognition. APOBEC1 has a preference for editing sites within AU rich sequences.[3] Moreover, the sequence specificity for apoB mRNA editing is provided by the 11 nucleotide long mooring sequence (UGAUCAGUAUA) that is located ~5 nucleotides downstream from the editing site. ACF binds the mooring sequence at the 3′ end of the edited C, thereby facilitating access to the editing site for APOBEC-1 (Fig. 1B). ACF contains three single-stranded RNA recognition motifs (RRMs) as well as a region with six RG residues and a COOH-terminal double-stranded RNA binding domain (dsRBD), which is not essential for editing.[4] RRM2 and RRM3 are the most important domains of ACF for RNA binding and APOBEC1 complementation function via protein-protein interactions. More precisely, ACF interacts with APOBEC1 via residues 144-257 (most part of RRM 2 + 3), and RNA binding is mainly accomplished by residues 150-380 (RRM 2 + 3 and part of the RG cluster).[5] However, all three RRMs are required for complementing activity and APOBEC1 interaction, only the different motifs account for RNA-binding to different extents. Additionally to the RRMs also N- and C-terminal amino acids flanking the RRMs are crucial for RNA binding.[7] APOBEC1 itself also contains a low RNA binding activity.[46] However, for functional editing ACF is indispensable and may help to allow access to the edited nucleotide by unwinding substrate RNA.[47] Recently it was shown that a truncated version of ACF (1-320), which is still able to bind RNA and APOBEC-1, self-associates in vivo by forming homo-multimers that are bridged by RNA. The authors speculate that ACF function might require a mooring-sequence-RNA bridged by head-to-head self-association.[48]

Recently, a screening for APOBEC1 editing targets by high throughput sequencing in mouse small intestine enterocytes identified several new editing sites. These sites are, unlike the apoB site, located in 3' UTRs of different transcripts. All of the sites are located in AU-rich regions and most of them have a motif similar to the mooring sequence downstream of the edited C. Additionally, the identified C to U editing sites lie within conserved regions of the respective 3' UTRs pointing to possible functional roles.[49]

PPR Proteins: Editing Factors in Plant Organelles

In plastids and mitochondria of plants the most abundant-type of editing is C to U editing, but U to C changes can sometimes also be found. Usually editing occurs in coding regions, where many C to U or U to C editing sites are important for proper protein function, as they create start or stop codons or change the amino acid sequence. Nonetheless, this type of editing is also found in some tRNAs.[49]

A cis-element neighboring the editing site specifies the site of editing by serving as a binding site for the plastid and mitochondrial editing factors.[50] To date the best understood factors required for RNA-editing in plant plastids and mitochondria are the so called pentatricopeptide repeat proteins (PPR proteins) due to their repeated motif of about 35 amino acids. Nearly all of them belong to the E subclass, as they have a C-terminal E-domain. However, recently also a P subfamily protein, PPR596, does influence editing efficiency at partial editing sites.[51] The PPR proteins are thought to serve as site specific accessory factors that bind to the RNA at the editing site. For one of the many PPR proteins, the plastid PPR protein CRR4, specific RNA binding to the surrounding of the editing site in ndhD-1 mRNA has been proven.[52] Many PPR proteins also have a second C-terminal extension after the E domain, which characteristically ends with a DYW triplet.[53] Interestingly, there is no general requirement for a DYW domain for the E-domain PPR proteins to perform editing, as also PPR proteins lacking this domain can function in editing.[54] This is also supported by the identification of five mitochondrial E-class PPR proteins involved in editing of which only one contains a C-terminal DYW domain.[55] Although the DYW domains contain amino acids typical for Zn-containing cytidine deaminases they have not been found to have cytidine deaminase activity. However, they have been shown to possess RNase activity that cleaves before adenosines in the RNA molecule.[53] Another feature of the family of PPR proteins is their apparent partial functional redundancy: Although some editing sites depend on one specific PPR protein, others can still be partially edited even if the corresponding PPR protein is missing.

A to I Modifications

Deamination of adenosines leads to the formation of inosines. Inosines can basepair with cytosines and therefore have the basepairing potential of guanosines.[56] Moreover, conversion of A-U basepairs to the less favorable I-U basepair conformation can also lead to the destabilization of structured and double stranded regions in RNA.

In fact, the deaminase activity responsible for the A to I conversion was originally described as an RNA unwinding activity in Xenopus oocytes.[57] In the meantime the enzymatic activity leading to the conversion of adenosines to inosines in double stranded RNA has been found in all metazoa tested.[58] Moreover, enzymes performing this reaction were cloned from several animals.[59,60] Depending on the organism, several proteins with deaminase activity have been cloned. These enzymes all belonging to the family of adenosine deaminases that act on RNAs (ADARs) have a very similar architecture harboring a C-terminal deaminase domain, a variable number of double-stranded RNA-binding domains and a rather heterogeneous N-terminus that can contain additional RNA or DNA-binding motifs.[61,62]

Besides this group of adenosine deaminases that act on RNA, a second group of related enzymes that act on tRNA is present in all kingdoms of life.[63]

ADARs

Adenosine deaminases that act on RNA can act both specific and nonspecific.[64-68] In specific editing a single or a few adenosines in an mRNA are converted to an inosine. Frequently codons are affected, leading to the translation of a protein that differs from its genomically encoded version. Since not all mRNAs are affected equally and in all tissues, this type of editing leads to proteomic diversification from a limited number of genes. To this point the number of known editing events that leads to amino acid exchanges is limited.[69] However, until recently, only editing events that occurred at relatively high abundance, thus affecting a significant proportion of all mRNAs in a given tissue could be detected. Recently, with the advent of novel sequencing technologies it also has become possible to detect rare editing events that affect only a small proportion of RNAs. Thus, next generation sequencing has just recently led to the discovery of numerous low abundance editing sites in the human transcriptome.[70]

Besides the specific editing events that lead to alteration in codons and thus to the formation of novel proteins or to the change of splice sites, a large number of editing events can be detected in inverted repetitive elements located throughout many genomes. Insertion of repetitive elements that primarily belong to the SINE family can frequently occur in inverted orientation in transcribed portions of genes. Such inverted repeats can then lead to the formation of double-stranded structures in a transcript. Double stranded structures are subsequently bound and edited by ADARs. The immediate consequences of this rather promiscuous type of editing are not clear. However, the presence of edited elements in untranslated regions has been linked to a reduction in gene expression both in the nematode *C. elegans* and in mammalian cells.[71,72]

A key question in both specific but also promiscuous RNA editing is how substrate RNAs and more specifically individual adenosines are being recognized and selected for editing. Depending on the organism and member of the ADAR family a variable number of double-stranded RNA binding domains are located in the center of the protein. dsRBDs were characterized as domains binding double-stranded A form RNA helices.[73,74] At least two helical turns are required to accommodate a dsRBD. Initial studies have mapped the interactions of the dsRBD to the backbone side of the RNA, measuring the distances of the characteristic A form and B form helices but also precluding sequence specific contacts. Subsequent studies have shown, however, that dsRBDs can bind cooperatively thereby helping to position each other.[75] Since naturally occurring helices are mostly of limited length and frequently interrupted by bulges, this can lead to a very specific positioning of adjacent dsRBDs.

Most interestingly, studies on a minimal stem-loop substrate for ADAR2, the R/G site in the glutamate receptor subunit B has shown that different dsRBDs can have both different structural but also sequence requirements for binding (see Chapter 9 by Cléry and Allain). It was shown that some dsRBDs can satisfy some of their binding sites with a terminal loop region.[76] This, in turn can help to a very specific positioning of a particular dsRBD. In the case of the two dsRBD containing ADAR2 protein, the interaction of one dsRBD with the terminal loop region leads to a very precise positioning of this dsRBD which in turn forces the other dsRBD quite precisely into place (Fig. 1C).[75]

Besides this structure based positioning of dsRBDs a sequence specific binding of dsRBDs has also been observed most recently.[77] In line with this finding, previous studies have already shown a specific interaction of some dsRBDs with specific transcriptionally active sites in vivo.[78] Thus, specific recognition of editing sites can both depend on structure and sequence.

Besides the double-stranded RNA binding domains making contact between ADARs and their substrate RNAs, other regions and domains in ADARs contribute to substrate binding and editing specificity.

First, mammalian ADAR1 contains a Z-DNA binding domain at its amino terminus. This domain may help to guide ADAR1 to transcriptionally active chromatin regions.[79] Alternatively, this region has been shown to allow interaction with left-handed RNA helices, thus also contributing to substrate interaction.[80]

Mammalian ADAR3 is a protein lacking enzymatic activity. Still, the RNA-binding domains and even the deaminase domain are largely conserved suggesting a function as an RNA binding protein, possibly as an antagonist or interaction partner of other ADARs. In agreement with this notion ADAR3 contains an additional arginine rich region in its amino terminus enhancing its interaction with single stranded RNA.[62]

Finally, substrate specificity of ADARs emerges from its deaminase domain.[81,82] The domain requires inositol hexakisphosphate for proper folding and coordintation of the zinc cofactor.[81] Moreover, the crystal structure of the domain explains why only adenosines but no other nucleotides fit into the catalytic domain. This unique fit could also be verified by mutagenesis studies in which the substrate specificity could be changed by mutagenesis of critical residues in the catalytic domain.[82]

ADATs

Modification and editing of tRNAs is widespread in all kingdoms of life including organellar tRNAs. ADATs are adenosine deaminases that act on tRNAs. A to I editing occurs predominantly at the wobble position of the anticodon, allowing translation of redundant codons.[83] However, additional A to I editing can occur in the position 37 of eukaryotic tRNA[Ala].[63]

The enzymes responsible for this type of editing vary in different organismic groups. In eukaryotes, a dimer between TAD2 (ADAT2) and TAD3 (ADAT3) is responsible for editing the wobble position, while the same position is edited by a tadA homodimer in prokaryotes.[84,85] In eukaryotes TAD1 (ADAT1) is responsible for editing of position 37 in tRNA[ala]. ADAT1 lacks any obvious RNA binding domains and thus, RNA recognition might be intrinsic to the deaminase domain.[63,86,87]

Structural and biochemical studies on bacterial tadA have shown that a minimal stem-loop RNA can be recognized by a tadA homodimer.[85] A C-terminal extension that is unique to bacteria tadA may also be involved in substrate binding.[88] In contrast, the eukaryotic ADAT2/ADAT3 heterodimer cannot recognize a minimal stem-loop RNA but rather requires a full-length tRNA for efficient editing. Thus, substrate recognition by the considerably larger eukaryotic ADAT2/3 heterodimer will clearly require additional interactions with the full-length tRNA.[84]

In plants tadA is rather unusual by showing a long N-terminal extension. Nonetheless, the function of this extension is currently unknown.[14]

Interestingly, the catalytic domain in ADATs resembles that of cytidine deaminases rather than that of adenosine deaminases. Consistently, in *T. brucei* the ADAT2/3 heterodimer is required for both A to I deamination at position 34 but also for a C to U deamination at position 32 of tRNA[thr].[89] Catalysis is most likely performed by the ADAT2 moiety. However, while the heterodimer is able to perform the A to I deamination reaction in vitro. The purified complex can only deaminate cytosines in DNA but not in RNA.[89]

Placozoa

Quite recently, a C to U editing event was found to occur in the Placozoan *Trichoplax adhaerens*, which is known to be one of the simplest free-living animals. Editing occurs in the *cox1* mRNA and converts a genomically encoded tyrosine UAU codon into an evolutionarily conserved CAU histidine codon. The resulting histidine is essential due to its importance for copper ion coordination.[90]

Conclusion

Over the past ten to 15 years, our view on the impact and significance of RNA editing has changed dramatically. On the one hand, the molecular machineries performing the editing reactions and substrate recognition have in many cases been characterized to atomic resolution. On the other hand advances in sequencing technology has proven RNA-editing to be tremendously widespread and abundant.

A challenge for the next years will be to understand the impact of RNA-editing events on the transcriptome and consequently on organismic life. More genomic and transcriptome sequence data has also lead to indications of novel, hitherto unknown RNA-editing events. Deciphering editing machineries that are responsible for these editing events will be another challenge of the upcoming years.

Acknowledgments

Work in the lab of MFJ is supported by the Austrian Science Foundation. CV is supported by the doctoral program in RNA-Biology, funded by the Austrian Science Foundation and the University of Vienna.

References

1. Anant S, MacGinnitie AJ, Davidson NO. apobec-1, the catalytic subunit of the mammalian apolipoprotein B mRNA editing enzyme, is a novel RNA-binding protein. J Biol Chem 1995; 270:14762-14767.
2. Davidson NO, Anant S, MacGinnitie AJ. Apolipoprotein B messenger RNA editing: insights into the molecular regulation of posttranscriptional cytidine deamination. Curr Opin Lipidol 1995; 6:70-74.
3. Anant S, MacGinnitie AJ, Davidson NO. The binding of apobec-1 to mammalian apo B RNA is stabilized by the presence of complementation factors which are required for posttranscriptional editing. Nucleic Acids Symp Ser 1995:99-102.
4. Mehta A, Kinter MT, Sherman NE et al. Molecular cloning of apobec-1 complementation factor, a novel RNA-binding protein involved in the editing of apolipoprotein B mRNA. Mol Cell Biol 2000; 20:1846-1854.
5. Blanc V, Henderson JO, Kennedy S et al. Mutagenesis of apobec-1 complementation factor reveals distinct domains that modulate RNA binding, protein-protein interaction with apobec-1, and complementation of C to U RNA-editing activity. J Biol Chem 2001; 276:46386-46393.
6. Blanc V, Navaratnam N, Henderson JO et al. Identification of GRY-RBP as an apolipoprotein B RNA-binding protein that interacts with both apobec-1 and apobec-1 complementation factor to modulate C to U editing. J Biol Chem 2001; 276:10272-10283.
7. Mehta A, Driscoll DM. Identification of domains in apobec-1 complementation factor required for RNA binding and apolipoprotein-B mRNA editing. RNA 2002; 8:69-82.
8. Hou Y, Lin S. Distinct gene number-genome size relationships for eukaryotes and non-eukaryotes: gene content estimation for dinoflagellate genomes. PLoS One 2009; 4:e6978.
9. Loya CM, Van Vactor D, Fulga TA. Understanding neuronal connectivity through the posttranscriptional toolkit. Genes Dev 2010; 24:625-635.
10. Blum B, Bakalara N, Simpson L. A model for RNA editing in kinetoplastid mitochondria: "guide" RNA molecules transcribed from maxicircle DNA provide the edited information. Cell 1990; 60:189-198.
11. Blum B, Simpson L. Guide RNAs in kinetoplastid mitochondria have a nonencoded 3' oligo(U) tail involved in recognition of the preedited region. Cell 1990; 62:391-397.
12. Benne R. RNA editing in trypanosomes. The us(e) of guide RNAs. Mol Biol Rep 1992; 16:217-227.
13. Aphasizhev R, Sbicego S, Peris M et al. Trypanosome mitochondrial 3' terminal uridylyl transferase (TUTase): the key enzyme in U-insertion/deletion RNA editing. Cell 2002; 108:637-648.
14. Aphasizhev R, Aphasizheva I, Nelson RE et al. Isolation of a U-insertion/deletion editing complex from Leishmania tarentolae mitochondria. EMBO J 2003; 22:913-924.
15. Hermann T, Schmid B, Heumann H et al. A three-dimensional working model for a guide RNA from Trypanosoma brucei. Nucleic Acids Res 1997; 25:2311-2318.
16. Niemann M, Kaibel H, Schluter E et al. Kinetoplastid RNA editing involves a 3' nucleotidyl phosphatase activity. Nucleic Acids Res 2009; 37:1897-1906.
17. Carnes J, Trotter JR, Peltan A et al. RNA editing in Trypanosoma brucei requires three different editosomes. Mol Cell Biol 2008; 28:122-130.
18. Worthey EA, Schnaufer A, Mian IS et al. Comparative analysis of editosome proteins in trypanosomatids. Nucleic Acids Res 2003; 31:6392-6408.
19. Schnaufer A, Wu M, Park YJ et al. A protein-protein interaction map of trypanosome ~20S editosomes. J Biol Chem 2010; 285:5282-5295.
20. Panigrahi AK, Schnaufer A, Ernst NL et al. Identification of novel components of Trypanosoma brucei editosomes. RNA 2003; 9:484-492.
21. Schumacher MA, Karamooz E, Zikova A et al. Crystal structures of T. brucei MRP1/MRP2 guide-RNA binding complex reveal RNA matchmaking mechanism. Cell 2006; 126:701-711.
22. Zikova A, Kopecna J, Schumacher MA et al. Structure and function of the native and recombinant mitochondrial MRP1/MRP2 complex from Trypanosoma brucei. Int J Parasitol 2008; 38:901-912.
23. Pelletier M, Read LK. RBP16 is a multifunctional gene regulatory protein involved in editing and stabilization of specific mitochondrial mRNAs in Trypanosoma brucei. RNA 2003; 9:457-468.
24. Hashimi H, Cicova Z, Novotna L et al. Kinetoplastid guide RNA biogenesis is dependent on subunits of the mitochondrial RNA binding complex 1 and mitochondrial RNA polymerase. RNA 2009; 15:588-599.
25. Hashimi H, Zikova A, Panigrahi AK et al. TbRGG1, an essential protein involved in kinetoplastid RNA metabolism that is associated with a novel multiprotein complex. RNA 2008; 14:970-980.
26. Mahendran R, Spottswood MS, Ghate A et al. Editing of the mitochondrial small subunit rRNA in Physarum polycephalum. EMBO J 1994; 13:232-240.
27. Gott JM, Parimi N, Bundschuh R. Discovery of new genes and deletion editing in Physarum mitochondria enabled by a novel algorithm for finding edited mRNAs. Nucleic Acids Res 2005; 33:5063-5072.
28. Gott JM, Visomirski LM, Hunter JL. Substitutional and insertional RNA editing of the cytochrome c oxidase subunit 1 mRNA of Physarum polycephalum. J Biol Chem 1993; 268:25483-25486.

29. Byrne EM, Gott JM. Cotranscriptional editing of Physarum mitochondrial RNA requires local features of the native template. RNA 2002; 8:1174-1185.
30. Vidal S, Curran J, Kolakofsky D. A stuttering model for paramyxovirus P mRNA editing. EMBO J 1990; 9:2017-2022.
31. Cheng YW, Visomirski-Robic LM, Gott JM. Non-templated addition of nucleotides to the 3′ end of nascent RNA during RNA editing in Physarum. EMBO J 2001; 20:1405-1414.
32. Miller ML, Miller DL. Non-DNA-templated addition of nucleotides to the 3′ end of RNAs by the mitochondrial RNA polymerase of Physarum polycephalum. Mol Cell Biol 2008; 28:5795-5802.
33. Rhee AC, Somerlot BH, Parimi N et al. Distinct roles for sequences upstream of and downstream from Physarum editing sites. RNA 2009; 15:1753-1765.
34. Gott JM, Somerlot BH, Gray MW. Two forms of RNA editing are required for tRNA maturation in Physarum mitochondria. RNA 2010; 16:482-488.
35. Lonergan KM, Gray MW. Editing of transfer RNAs in Acanthamoeba castellanii mitochondria. Science 1993; 259:812-816.
36. Laforest MJ, Roewer I, Lang BF. Mitochondrial tRNAs in the lower fungus Spizellomyces punctatus: tRNA editing and UAG 'stop' codons recognized as leucine. Nucleic Acids Res 1997; 25:626-632.
37. Forget L, Ustinova J, Wang Z et al. Hyaloraphidium curvatum: a linear mitochondrial genome, tRNA editing, and an evolutionary link to lower fungi. Mol Biol Evol 2002; 19:310-319.
38. Price DH, Gray MW. A novel nucleotide incorporation activity implicated in the editing of mitochondrial transfer RNAs in Acanthamoeba castellanii. RNA 1999; 5:302-317.
39. Bullerwell CE, Gray MW. In vitro characterization of a tRNA editing activity in the mitochondria of Spizellomyces punctatus, a Chytridiomycete fungus. J Biol Chem 2005; 280:2463-2470.
40. Zauner S, Greilinger D, Laatsch T et al. Substitutional editing of transcripts from genes of cyanobacterial origin in the dinoflagellate Ceratium horridum. FEBS Lett 2004; 577:535-538.
41. Nash EA, Barbrook AC, Edwards-Stuart RK et al. Organization of the mitochondrial genome in the dinoflagellate Amphidinium carterae. Mol Biol Evol 2007; 24:1528-1536.
42. Smith HC, Gott JM, Hanson MR. A guide to RNA editing. RNA 1997; 3:1105-1123.
43. Pak BJ, Pang SC. Developmental regulation of the translational repressor NAT1 during cardiac development. J Mol Cell Cardiol 1999; 31:1717-1724.
44. Mukhopadhyay D, Anant S, Lee RM et al. C—>U editing of neurofibromatosis 1 mRNA occurs in tumors that express both the type II transcript and apobec-1, the catalytic subunit of the apolipoprotein B mRNA-editing enzyme. Am J Hum Genet 2002; 70:38-50.
45. Chen KM, Harjes E, Gross PJ et al. Structure of the DNA deaminase domain of the HIV-1 restriction factor APOBEC3G. Nature 2008; 452:116-119.
46. Navaratnam N, Bhattacharya S, Fujino T et al. Evolutionary origins of apoB mRNA editing: catalysis by a cytidine deaminase that has acquired a novel RNA-binding motif at its active site. Cell 1995; 81:187-195.
47. Maris C, Masse J, Chester A et al. NMR structure of the apoB mRNA stem-loop and its interaction with the C to U editing APOBEC1 complementary factor. RNA 2005; 11:173-186.
48. Galloway CA, Kumar A, Krucinska J et al. APOBEC-1 complementation factor (ACF) forms RNA-dependent multimers. Biochem Biophys Res Commun 2010; 398:38-43.
49. Shikanai T. RNA editing in plant organelles: machinery, physiological function and evolution. Cell Mol Life Sci 2006; 63:698-708.
50. Takenaka M, Verbitskiy D, van der Merwe JA et al. The process of RNA editing in plant mitochondria. Mitochondrion 2008; 8:35-46.
51. Doniwa Y, Ueda M, Ueta M et al. The involvement of a PPR protein of the P subfamily in partial RNA editing of an Arabidopsis mitochondrial transcript. Gene 2010; 454:39-46.
52. Okuda K, Nakamura T, Sugita M et al. A pentatricopeptide repeat protein is a site recognition factor in chloroplast RNA editing. Journal of Biological Chemistry 2006; 281:37661-37667.
53. Nakamura T, Sugita M. A conserved DYW domain of the pentatricopeptide repeat protein possesses a novel endoribonuclease activity. FEBS Lett 2008; 582:4163-4168.
54. Okuda K, Chateigner-Boutin AL, Nakamura T et al. Pentatricopeptide repeat proteins with the DYW motif have distinct molecular functions in RNA editing and RNA cleavage in Arabidopsis chloroplasts. Plant Cell 2009; 21:146-156.
55. Takenaka M, Verbitskiy D, Zehrmann A et al. Reverse genetic screening identifies five E-class PPR proteins involved in RNA editing in mitochondria of Arabidopsis thaliana. J Biol Chem 2010; 285:27122-27129.
56. Polson AG, Crain PF, Pomerantz SC et al. The mechanism of adenosine to inosine conversion by the double-stranded RNA unwinding/modifying activity: a high-performance liquid chromatography-mass spectrometry analysis. Biochemistry 1991; 30:11507-11514.

57. Bass BL, Weintraub H. A developmentally regulated activity that unwinds RNA duplexes. Cell 1987; 48:607-613.
58. Wagner RW, Yoo C, Wrabetz L et al. Double-stranded RNA unwinding and modifying activity is detected ubiquitously in primary tissues and cell lines. Mol Cell Biol 1990; 10:5586-5590.
59. O'Connell MA, Krause S, Higuchi M et al. Cloning of cDNAs encoding mammalian double-stranded RNA-specific adenosine deaminase. Mol Cell Biol 1995; 15:1389-1397.
60. Kim U, Wang Y, Sanford T et al. Molecular cloning of cDNA for double-stranded RNA adenosine deaminase, a candidate enzyme for nuclear RNA editing. Proc Natl Acad Sci U S A 1994; 91:11457-11461.
61. Herbert A, Alfken J, Kim YG et al. A Z-DNA binding domain present in the human editing enzyme, double- stranded RNA adenosine deaminase. Proc Natl Acad Sci U S A 1997; 94:8421-8426.
62. Chen CX, Cho DS, Wang Q et al. A third member of the RNA-specific adenosine deaminase gene family, ADAR3, contains both single- and double-stranded RNA binding domains. RNA 2000; 6:755-767.
63. Gerber A, Grosjean H, Melcher T et al. Tad1p, a yeast tRNA-specific adenosine deaminase, is related to the mammalian pre-mRNA editing enzymes ADAR1 and ADAR2. EMBO J 1998; 17:4780-4789.
64. Burns CM, Chu H, Rueter SM et al. Regulation of serotonin-2C receptor G-protein coupling by RNA editing. Nature 1997; 387:303-308.
65. Sommer B, Kohler M, Sprengel R et al. RNA editing in brain controls a determinant of ion flow in glutamate-gated channels. Cell 1991; 67:11-19.
66. Kim DD, Kim TT, Walsh T et al. Widespread RNA editing of embedded alu elements in the human transcriptome. Genome Res 2004; 14:1719-1725.
67. Athanasiadis A, Rich A, Maas S. Widespread A-to-I RNA Editing of Alu-Containing mRNAs in the Human Transcriptome. PLoS Biol 2004; 2:e391.
68. Levanon EY, Eisenberg E, Yelin R et al. Systematic identification of abundant A-to-I editing sites in the human transcriptome. Nat Biotechnol 2004; 22:1001-1005.
69. Levanon EY, Hallegger M, Kinar Y et al. Evolutionarily conserved human targets of adenosine to inosine RNA editing. Nucleic Acids Res 2005; 33:1162-1168.
70. Li JB, Levanon EY, Yoon JK et al. Genome-wide identification of human RNA editing sites by parallel DNA capturing and sequencing. Science 2009; 324:1210-1213.
71. Hundley HA, Krauchuk AA, Bass BL. C. elegans and H. sapiens mRNAs with edited 3' UTRs are present on polysomes. RNA 2008; 14:2050-2060.
72. Chen LL, DeCerbo JN, Carmichael GG. Alu element-mediated gene silencing. EMBO J 2008; 27:1694-1705.
73. Ryter JM, Schultz SC. Molecular basis of double-stranded RNA-protein interactions: structure of a dsRNA-binding domain complexed with dsRNA. EMBO J 1998; 17:7505-7513.
74. Ramos A, Grunert S, Adams J et al. RNA recognition by a Staufen double-stranded RNA-binding domain. EMBO J 2000; 19:997-1009.
75. Stefl R, Xu M, Skrisovska L et al. Structure and specific RNA binding of ADAR2 double-stranded RNA binding motifs. Structure 2006; 14:345-355.
76. Stefl R, Allain FH. A novel RNA pentaloop fold involved in targeting ADAR2. RNA 2005; 11:592-597.
77. Stefl R, Oberstrass FC, Hood JL et al. The solution structure of the ADAR2 dsRBM-RNA complex reveals a sequence-specific readout of the minor groove. Cell 2010; 143:225-237.
78. Doyle M, Jantsch MF. Distinct in vivo roles for double-stranded RNA-binding domains of the Xenopus RNA-editing enzyme ADAR1 in chromosomal targeting. The Journal of cell biology 2003; 161:309-319.
79. Herbert A, Schade M, Lowenhaupt K et al. The Zalpha domain from human ADAR1 binds to the Z-DNA conformer of many different sequences. Nucleic Acids Res 1998; 26:3486-3493.
80. Brown BA 2nd, Lowenhaupt K, Wilbert CM et al. The zalpha domain of the editing enzyme dsRNA adenosine deaminase binds left-handed Z-RNA as well as Z-DNA. Proc Natl Acad Sci U S A 2000; 97:13532-13536.
81. Macbeth MR, Schubert HL, Vandemark AP et al. Inositol hexakisphosphate is bound in the ADAR2 core and required for RNA editing. Science 2005; 309:1534-1539.
82. Pokharel S, Jayalath P, Maydanovych O et al. Matching active site and substrate structures for an RNA editing reaction. J Am Chem Soc 2009; 131:11882-11891.
83. Gerber AP, Keller W. RNA editing by base deamination: more enzymes, more targets, new mysteries. Trends Biochem Sci 2001; 26:376-384.
84. Gerber AP, Keller W. An adenosine deaminase that generates inosine at the wobble position of tRNAs. Science 1999; 286:1146-1149.
85. Wolf J, Gerber AP, Keller W. tadA, an essential tRNA-specific adenosine deaminase from Escherichia coli. EMBO J 2002; 21:3841-3851.
86. Maas S, Gerber AP, Rich A. Identification and characterization of a human tRNA-specific adenosine deaminase related to the ADAR family of pre-mRNA editing enzymes. Proc Natl Acad Sci U S A 1999; 96:8895-8900.

87. Keegan LP, Gerber AP, Brindle J et al. The properties of a tRNA-specific adenosine deaminase from Drosophila melanogaster support an evolutionary link between pre-mRNA editing and tRNA modification. Mol Cell Biol 2000; 20:825-833.

88. Kuratani M, Ishii R, Bessho Y et al. Crystal structure of tRNA adenosine deaminase (TadA) from Aquifex aeolicus. J Biol Chem 2005; 280:16002-16008.

89. Rubio MA, Pastar I, Gaston KW et al. An adenosine-to-inosine tRNA-editing enzyme that can perform C-to-U deamination of DNA. Proc Natl Acad Sci U S A 2007; 104:7821-7826.

90. Burger G, Yan Y, Javadi P et al. Group I-intron trans-splicing and mRNA editing in the mitochondria of placozoan animals. Trends Genet 2009; 25:381-386.

CHAPTER 8

RNA-Binding Proteins in Plant Response to Abiotic Stresses

Hunseung Kang and Kyung Jin Kwak*

Abstract

RNA-binding proteins (RBPs) are recognized as key regulatory factors in the posttranscriptional regulation of gene expression in eukaryotes. Over the past decades, considerable progress has been made in the identification of RBPs active in the stress response as well as in development and growth of plants. Despite the fact that RBPs have been implicated to play roles in plant responses to diverse environmental stresses, determination of their functional roles under stress conditions lags far behind their identification in various plant species. However, recent advances in functional characterization of several RBP families have shed light on the functional roles and action mechanisms of RBPs in the plant response to changing environmental conditions. In particular, recent reports have demonstrated the emerging idea that some RBPs act as RNA chaperones during the stress adaptation process in both monocotyledonous plants and dicotyledonous plants.

Introduction

The regulation of gene expression in living organisms occurs at the posttranscriptional level as well as transcriptional level, and both are of crucial importance for growth, development and stress responses. In posttranscriptional gene regulatory mechanisms such as pre-mRNA splicing, capping, polyadenylation, mRNA transport, stability and translation,[1-3] regulation is mainly achieved either directly by RNA-binding proteins (RBPs) or indirectly, whereby RBPs modulate the function of other regulatory factors. The question of how a RBP recognizes a specific RNA site and promotes a specific RNA function is thus the key to understanding many cellular processes.

RNA-binding proteins as well as RNA-protein complexes have been investigated in a variety of living organisms, including microorganisms, animals and plants. This has led to the discovery of several conserved protein motifs, such as RNA-recognition motifs (RRMs), glycine-rich domains, arginine-rich domains, SR-repeats, RD-repeats and zinc finger motifs.[4-8] The RRM is the most widely found and best characterized RNA-binding motif. RNA-binding proteins containing RRMs at their N-terminus and a glycine-rich region at their C-terminus, thus referred to as glycine-rich RNA-binding proteins (GRPs), have been implicated in plant responses to environmental stimuli, and some progress has been made to uncover the roles of GRPs in the responses of plants to changing environmental conditions.

Cold shock proteins (CSPs), which are highly induced during the cold acclimation phase in bacteria and fungi,[9-14] are homologous to the cold shock domain (CSD) of eukaryotic Y-box proteins, which are involved in the regulation of gene expression.[15] It has been suggested that certain

*Department of Plant Biotechnology, College of Agriculture and Life Sciences, Chonnam National University, Gwangju, Korea.
Corresponding Author: Hunseung Kang—Email: hskang@jnu.ac.kr

RNA Binding Proteins, edited by Zdravko J. Lorković.

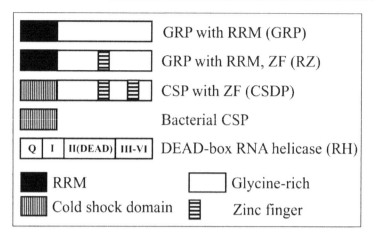

Figure 1. Schematic presentation of the domain structures of RNA-binding proteins discussed in this chapter. Glycine-rich RNA-binding proteins (GRPs) contain an RNA-recognition motif (RRM) at their N-terminus and a glycine-rich region at their C-terminus. RZs harbor an RRM at their N-terminus and a glycine-rich region interspersed with a CCHC-type zinc finger at their C-terminus. Cold shock domain proteins (CSDPs) contain a cold shock domain (CSD) in their N-terminal half as well as additional glycine-rich regions interspersed by CCHC-type zinc fingers in their C-terminal half. DEAD-box RNA helicases (RHs) contain motifs called Q, I, II (DEAD), III, IV, V and VI domains.

CSPs in *Escherichia coli* function as RNA chaperones that facilitate translation at low temperature by destabilizing secondary structures in mRNA.[16] Eukaryotes including plants also contain proteins harboring the CSD in their N-terminal half as well as additional glycine-rich regions interspersed with CCHC-type zinc fingers in their C-terminal half, thereby called cold shock domain proteins (CSDPs).[17] Although CSDPs have been implicated to play roles in the responses of plants to environmental stresses, the functional roles of CSDPs have not well been established.

RNA helicases (RHs) are ubiquitous enzymes that catalyze the unwinding of energetically stable duplex RNA secondary structures in an ATP-dependent manner, and play important roles in many cellular processes, including transcription, translation and RNA metabolism.[18] DEAD-box RNA helicases have been implicated to play roles during stress adaptation processes in plants. Since RNA molecules are prone to forming stable, nonfunctional secondary structures, their proper functioning requires unwinding of misfolded RNA molecules. RNA helicases are also potential candidates for RNA chaperones, which can disrupt misfolded RNA structures and mediate correct folding.[19,20] The roles of RHs as RNA chaperones are more prominent when cells are exposed to low temperatures, as misfolded RNA molecules become stabilized and cannot assume native conformation without the help of RNA chaperones. However, the functions of RHs in the responses of plants to environmental stimuli are poorly understood.[21] Given that GRPs, CSDPs and RHs all have the potential to function as regulators of gene expression during the stress adaptation process in plants, this chapter will discuss the latest research concerning the structural and functional aspects of GRPs, CSDPs and RHs (Fig. 1) in abiotic stress responses, and particularly their roles as RNA chaperones during cold adaptation in plants.

Structural Diversity of Plant RNA-Binding Proteins
RRM-harboring RBPs contain one or more RRMs at their N-terminus as well as auxiliary motifs, such as glycine-rich domains, arginine-rich domains, SR-repeats, RD-repeats and zinc finger motifs at their C-terminus.[4,5,7] The RRM is the most widely abundant and best characterized RNA-binding motif. It comprises 80-90 amino acids and is present in one or multiple copies in proteins capable

of binding to pre-mRNAs, mRNAs and small nuclear RNAs. The most conserved RRM sequences are ribonucleoprotein 1 (RNP1) (eight amino acids; RGFGFVTF) and RNP2 (six amino acids; CFVGGL), which recognize and bind target RNA molecules. The conserved RNP1 and RNP2 sequences are juxtaposed on two central β strands, with the side chains of the conserved aromatic amino acids of RNP1 and RNP2 displayed on the surface of the β sheet (see Chapter 9 by Cléry and Allain for details). The auxiliary domains are observed at the C-terminus, which contribute to RNA-binding specificity.[4,22,23] Many RBPs contain multiple copies of the RRM, which normally determines the affinity for RNA substrates and the specificity of the proteins.

Plant glycine-rich proteins are characterized by the presence of repetitive glycine-rich motifs (see Chapter 3 by Rogelj et al for functions of glycine-rich domains) and they are classified based on the arrangement of glycine repeats and the presence of conserved motifs.[24] The GRPs discussed in this chapter contain an RRM at their N-terminus and a glycine-rich region at their C-terminus, which bind a variety of RNA molecules and participate at several levels of RNA metabolism and processing. Plant GRPs are small in size (approximately 16-20 kDa) and consist of two very diverse regions. The RRM occupies the N-terminal half of the protein and shows the highest sequence homology among all proteins (60-80% identity). The C-terminal region is rich in glycine residues (about 70%) and arginine residues (10-15%) and plays a role in the protein-protein interactions required for overall function. The *Arabidopsis* genome contains eight GRPs[7] while rice genome harbors at least six GRPs.[25] The amino acid sequences of GRPs in *Arabidopsis thaliana* and rice (*Oryza sativa*) are homologous to each other. In addition to these GRPs, RBPs harboring an RRM at their N-terminus and a glycine-rich region interspersed with a CCHC-type zinc finger at their C-terminus, designated as RZs (Fig. 1), have been identified in *Arabidopsis* and rice.[7,26] The *Arabidopsis* and rice genomes contain a gene family encoding three RZ proteins. The amino acid sequences of AtRZs and OsRZs share a sequence homology of approximately 40% with each other, and the position of the CCHC-type zinc finger motif is highly conserved in all RZ proteins.[26]

Cold shock proteins are homologous to a domain called the CSD of eukaryotic Y-box proteins.[27] The CSD is a highly conserved nucleic acid-binding domain comprising approximately 65 to 75 amino acid residues and is capable of binding RNA, single-stranded DNA and double-stranded DNA.[28-30] The structure of the CSD belongs to the oligosaccharide-/oligonucleotide-binding (OB) fold and contains conserved RNP1 and RNP2 motifs. Although unrelated in primary sequence, CSDs are structurally and functionally similar to S1 domains that are found in bacterial ribosomal S1 protein.[31] Structure determination revealed that the CSDs of bacterial cold-shock protein A, human Y-box binding protein and human upstream of N-ras protein fold into five-stranded anti-parallel β-barrels.[32-34] Contrary to the small size (about 7-10 kDa) of CSPs found in prokaryotes, plant CSDPs contain a CSD in their N-terminal half as well as additional glycine-rich regions interspersed by CCHC-type zinc fingers in their C-terminal half (Fig. 1).[35-38] It has been determined that *Arabidopsis* and rice genomes contain four CSDPs that have highly homologous CSDs in their N-terminal half but variable glycine-rich regions interspersed with different numbers of CCHC-type zinc fingers in their C-terminal half.[37,39]

RNA helicases are divided into six main groups, named superfamily (SF) 1 to SF6,[40,41] and all contain nine so-called helicase motifs, Q, I, Ia, Ib, II, III, IV, V and VI domains. Motifs I and II are highly conserved Walker A (ATPase I) and B (ATPase II) sequences characteristic of ATPases.[42] Helicases exhibit nucleic acid-dependent ATPase activity, which provide energy for the helicase action.[43-45] Some of these proteins are able to unwind double-stranded RNA as well as DNA-RNA hybrids, indicating an essential role for helicases in RNA metabolism. RNA helicases catalyze the ATP-dependent unwinding of local RNA secondary structures and play a role in altering RNA structures.[40,46,47] The DEAD-box family belongs to the SF2 helicase superfamily, the largest helicase superfamily, which contains a core region of highly conserved helicase and Q motifs.[20,40,48] The structures of several DEAD-box proteins have been solved in which the N-terminal domain and the C-terminal domain each conserve their folds, but the interdomain orientations are strikingly diverse.[49-52] The *Arabidopsis* genome harbors 58 DEAD-box RHs,[21,53,54] and the rice genome contains more than 50 DEAD-box RHs.[55]

Roles of RNA-Binding Proteins in Plant Stress Responses

Since the first gene encoding GRP was identified in maize,[56] cDNA encoding homologous proteins have been found in diverse plant species, including alfalfa, Arabidopsis, barley, Brassica, rice and tobacco.[57-67] The biological functions of plant GRPs in the responses of plants to changing environmental conditions have been investigated during the last decades, and the involvement of GRPs in diverse biological and biochemical processes are being uncovered.[24,68,69] In particular, the functions of plant GRPs during cold acclimation were elucidated based on the fact that they are significantly induced by low temperature. The biological roles of several GRPs in plants under stress conditions have been characterized recently (Table 1). It was reported that AtGRP2, AtGRP4 and AtGRP7, three AtGRPs of the eight AtGRP family members in Arabidopsis, have different impacts on seed germination, seedling growth and stress tolerance of Arabidopsis plants under diverse stress conditions.[70-72] In particular, AtGRP2 and AtGRP7 but not AtGRP4 accelerate seed germination and seedling growth under low temperatures and confer freezing tolerance to Arabidopsis plants. The results clearly reflect the prominent roles of GRPs in the responses of plants to diverse environmental factors in dicotyledonous plants such as Arabidopsis. Despite increasing knowledge regarding the functional roles of GRPs in Arabidopsis, our current understanding of their biological functions in monocotyledonous plants is severely limited. Proteomic analysis of RBPs revealed their existence and regulation in dry seeds of rice.[73] A recent study also showed that rice OsGRP1 and OsGRP4 confer cold and freezing tolerance in Arabidopsis plants,[25] demonstrating that GRPs in rice and Arabidopsis are functionally conserved during the cold adaptation process.

The biological roles of RZ proteins harboring glycine-rich regions interspersed with a CCHC-type zinc finger have been determined in limited cases (Table 1). Arabidopsis AtRZ-1a positively affects seed germination and seedling growth under cold stress conditions and contributes to the enhancement of freezing tolerance in Arabidopsis.[74,75] In contrast, AtRZ-1a has a negative impact on seed germination and seedling growth of Arabidopsis under salt or drought stress conditions.[76] Despite an increase in the transcription of AtRZ-1b and AtRZ-1c under cold stress, overexpression or loss-of-function mutations of this gene does not affect seed germination or seedling growth of Arabidopsis under cold stress conditions.[77] These results demonstrate that the three AtRZ-1 family members contribute differently to the enhancement of cold tolerance in plants. A recent report also has demonstrated that among the three RZ proteins in rice, OsRZ2 confer cold and freezing tolerance in Arabidopsis plants.[26]

Cold shock domain proteins have been implicated in the responses of plants to cold stress based on their cold-regulated expression patterns and their structural similarity to prokaryotic CSPs, which function as RNA chaperones for efficient translation at low temperatures.[16,78] Although the nucleic acid-binding properties and functions of bacterial CSPs have been well established, the functional roles of plant CSDPs have been determined only in limited cases. Arabidopsis AtCSDP2 (also designated AtGRP2) plays a role in flower and seed development,[79] and cold-regulated AtCSDP2 has RNA chaperone activity in E. coli.[79,80] The CSDPs isolated from winter wheat[81] and rice[39] harbor RNA chaperone activity during the cold adaptation process in E. coli. Arabidopsis AtCSDP3 has been determined to confer freezing tolerance,[82] and overexpression of AtCSDP1 delays seed germination of Arabidopsis under dehydration or salt stress conditions. On the other hand, AtCSDP2 overexpression accelerates seed germination of Arabidopsis under salt stress conditions. Moreover, AtCSDP1 and AtCSDP2 confer freezing tolerance in Arabidopsis.[83] In addition to their roles in stress response, AtCSDPs are suggested to be involved in flower and seed development in Arabidopsis.[79,80,84]

The roles of RHs in cellular processes such as ribosome biogenesis, pre-mRNA splicing, nucleo-cytoplasmic transport, translation and RNA decay have been established.[18] However, the roles of RHs in stress responses are largely unknown. The genes encoding DEAD-box RH, including csdA in E. coli, crhC in cyanobacteria and deaD in Methanococcoides burtonii, are induced in response to low temperatures.[19,85,86] Although it has been demonstrated that low

Table 1. Overview of plant RNA-binding proteins mentioned in the text and their roles in stress response

Protein Name	Plant Source	Gene ID	Expression/Function	References
Glycine-rich RNA-binding protein (GRP)				
AtGRP2	*Arabidopsis*	At4g13850	Cold/freezing tolerance	71,103
AtGRP4	*Arabidopsis*	At3g23830		70,103
AtGRP7	*Arabidopsis*	At2g21660	Cold/freezing tolerance	72,103
			Negative role to salt/drought	72
OsGRP1	Rice	Os01g68790	Cold/freezing tolerance	25
OsGRP4	Rice	Os04g33810	Cold/freezing tolerance	25
OsGRP6	Rice	Os12g31800	Cold/freezing tolerance	25
Zinc finger GRP (RZ)				
AtRZ-1a	*Arabidopsis*	At3g26420	Cold/freezing tolerance	74,75
			Negative role to salt/drought	76
AtRZ-1b	*Arabidopsis*	At1g60650		77
AtRZ-1c	*Arabidopsis*	At5g04280		77
OsRZ-1	Rice	Os03g61990		26
OsRZ-2	Rice	Os07g08960	Cold/freezing tolerance	26
OsRZ-3	Rice	Os03g47800		26
Cold shock domain protein (CSDP)				
AtCSDP1	*Arabidopsis*	At4g36020	Cold/freezing tolerance	103,104
			Negative role to salt/drought	83
AtCSDP2	*Arabidopsis*	At4g38680	Responds to cold	80,103,104
			Positive role to salt	83
AtCSDP3	*Arabidopsis*	At2g17870	Cold/freezing tolerance	82
AtCSDP4	*Arabidopsis*	At2g21060		
OsCSDP1	Rice	Os02g02870	Responds to cold	39
OsCSDP2	Rice	Os08g03520	Responds to cold	39
DEAD-box RNA helicase (RH)				
AtRH9	*Arabidopsis*	At3g22310	Cold/freezing tolerance	93
AtRH25	*Arabidopsis*	At5g08620	Cold/freezing tolerance	93
LOS4	*Arabidopsis*	At3g53110	Cold/freezing tolerance	90
STRS1	*Arabidopsis*	At1g31970	Suppressor of salt, osmotic, heat	92
STRS2	*Arabidopsis*	At5g08620	Suppressor of salt, osmotic, heat	92

temperature and other stresses regulate the expression of a diverse array of genes, including RHs in plants,[87,88] the number of reports addressing the functions of RHs in plant stress responses is severely limited. It was reported that low expression of an osmotically responsive gene (los4) encoding *Arabidopsis* RH (AtRH38) confers freezing tolerance.[89-91] The two DEAD-box RHs, STRESS RESPONSE SUPPRESSOR1 and STRESS RESPONSE SUPPRESSOR2, have been determined to attenuate *Arabidopsis* responses to multiple abiotic stresses.[92] Overexpression of *Arabidopsis* AtRH9 or AtRH25 results in retarded seed germination of *Arabidopsis* plants under salt stress conditions, whereas AtRH25 but not AtRH9 enhances freezing tolerance.[93] In addition to their roles in the cold stress response, DEAD-box RHs have also been implicated in other stress responses, including salt stress in barley,[94] oxidative stress in *Clostridium perfringens* and rice[95,96] and heavy metal stress in yeast.[97]

Molecular Mechanisms of RNA-Binding Proteins in Stress Adaptation Processes in Plants

Although increasing numbers of recent reports have clearly demonstrated that GRPs, CSDPs and RHs play diverse roles in the responses of plants to changing environmental conditions, the molecular mechanisms underlying these RBP-mediated stress responses in plants are just beginning to be uncovered. The most plausible mechanism by which these RBPs mediate their activity is the regulation of RNA folding in cells. RNA molecules have the tendency to fold into alternative secondary structures.[98] These alternative misfolded structures can then interfere with the normal function of RNA molecules and thus have to be resolved. It has been suggested that formation of alternative misfolded structures is prevented or reversed by the action of proteins generally known as RNA chaperones. The roles of RNA chaperones are more prominent when cells are exposed to low temperatures, as misfolded RNA molecules become stabilized and cannot assume native conformation without the help of RNA chaperones. It was determined that bacterial CSPs function as RNA chaperones by destabilizing the over-stabilized secondary structures in mRNAs for efficient translation at low temperatures.[12,16,99-101] Since it was found that cyanobacteria lack CSPs but contain a cold-induced RRM protein instead, it was hypothesized that RRM proteins may actually substitute for the function of CSPs in cyanobacteria.[29,102] Indeed, GRPs do contain similar structural features to RRM proteins found in cyanobacteria. Therefore, it is highly likely that GRPs as well as CSDPs function as RNA chaperones during the cold adaptation process in plants. This consideration is supported by a series of recent findings that demonstrated that GRPs, CSDPs and RHs harbor RNA chaperone activity. It was determined that *Arabidopsis* AtGRP2 and AtGRP7, which confer cold and freezing tolerance in plants,[71,72] complement the cold sensitivity of BX04 mutant *E. coli* cell, which lacks four CSPs and is highly sensitive to cold stress.[103] These AtGRPs contain nucleic acid-melting activities in vitro, which further demonstrates that GRPs have RNA chaperone activity during the cold adaptation process. In contrast, AtGRP4 does not increase the cold or freezing resistance of *Arabidopsis* plants[70] and does not have RNA chaperone activity.[103] Rice OsGRP1 and OsGRP4, which have the ability to confer cold and freezing tolerance in *Arabidopsis*, complement the cold-sensitive phenotype of BX04 mutant *E. coli* cells at low temperatures and harbor nucleic acid-melting ability.[25] This demonstrates that these rice GRPs exhibit RNA chaperone function during the cold adaptation process. In addition, *Arabidopsis* AtRZ-1a and rice OsRZ2, both of which contribute to the enhancement of freezing tolerance in *Arabidopsis*, have been determined to harbor RNA chaperone activity.[26,74,75] These findings strongly suggest that GRPs function as RNA chaperones during the cold adaptation process in plants as well as in bacteria.

The RNA chaperone activities of several CSDPs and RHs have also been determined. *Arabidopsis* AtCSDP1 comprising 299 amino acids with seven CCHC-type zinc fingers at the C-terminus possesses RNA chaperone activity, whereas AtCSDP2 comprising 204 amino acids with two CCHC-type zinc fingers at the C-terminus does not have RNA chaperone activity.[103] Domain-swapping and deletion experiments have shown that, although the CSD itself harbors RNA chaperone activity, the number and length of zinc finger glycine-rich domains of CSDPs are crucial to the full activity of the RNA chaperones.[104] The CSDPs isolated from winter wheat and rice also harbor RNA chaperone activity during the cold adaptation process in *E. coli*.[39,81] It has been determined that DEAD-box RHs are RNA chaperones that can disrupt misfolded RNA structures.[105,106] The *Arabidopsis* AtRH25, which confers enhanced freezing tolerance in *Arabidopsis* plants, also complements the cold-sensitive phenotype of BX04 *E. coli* mutant cells at low temperatures.[93] These results also indicate that CSDPs and RHs function as RNA chaperones during the cold adaptation process in bacteria and plants.

RNA-binding proteins exert their roles by binding to target RNAs and regulating RNA biogenesis and processing. Although the RRM and glycine-rich domain have been suggested to be involved in target RNA binding as well as in interactions with other proteins or ligand molecules,[107] the sequence-specific RNA-binding of plant GRPs has been analyzed in only a few cases. The GRPs from barley, maize and tobacco show high affinity for poly(G) and poly(U), indicating that the cellular target RNA is likely to be enriched in G- and U-residues.[108-111] *Arabidopsis* AtGRP2 binds preferentially to U-rich RNA sequences,[112] and AtGRP7 prefers to bind G- and U-rich RNA sequences.[103] In comparison, AtGRP4 binds nonspecifically to RNA homopolymers.[70] Recent microarray analysis of transcription upon AtGRP7 overexpression revealed that the transcripts associated with responses to abiotic or biotic stimuli are prevalently regulated by AtGRP7.[113] It was also observed that *E. coli* CspA binds single-stranded DNA and RNA sequences in a nonspecific manner,[16] and *Arabidopsis* AtCSDP1 binds preferentially to single-stranded DNA and G-rich RNA.[103] Although GRPs and CSDPs prefer to bind G- or U-rich RNA, the primary sequence of the RNA is not important.[103] The nonspecific nucleic acid-binding properties of GRPs and CSDPs reflect their roles as RNA chaperones, which generally bind RNAs sequence nonspecifically.

Localization studies have provided some clues as to the possible cellular roles of RBPs in posttranscriptional RNA metabolism. To function as an RNA chaperone that modulates RNA processing and metabolism, GRPs, CSDPs and RHs should be localized to the nucleus and/or the cytoplasm, where RNA processing and translation occurs. It has been determined that MA16 from maize and GRP from tobacco accumulate in the nucleolus or nucleoplasm, which suggests their participation in pre-ribosomal RNA processing or pre-mRNA metabolism.[62,114] It has recently been determined that *Arabidopsis* AtGRP7 is localized to the nucleus and cytoplasm,[72] and rice OsGRPs were determined to be localized to the nucleus.[25] Most DEAD box RHs are known to be localized to the nucleus as well. One possible activity of nuclear-localized GRPs and RHs is the regulation of mRNA export from the nucleus to the cytoplasm. It has been proposed that tobacco RZ-1 could be involved in pre-mRNA processing and/or nucleo-cytoplasmic transport.[107] Indeed, *Arabidopsis* GRPs and RHs have been determined to regulate mRNA export from the nucleus to the cytoplasm.[72,90,107,115] These results support the emerging idea that certain GRPs, CSDPs and RHs regulate mRNA export from the nucleus to the cytoplasm and thereby play a key role during the adaptation of plants to environmental stresses.

Conclusion and Future Prospects

Although the list of RBPs in diverse plant species is growing, the biological processes in which they participate are still poorly understood. In particular, GRPs are abundant plant proteins that appear to be involved in the plant responses to changing environmental conditions, yet their functions in this molecular process are largely unknown. Although our knowledge of the cellular functions of GRPs, CSDPs and RHs in stress responses remains far from sufficient, recent research on the conserved functions of specific GRPs, CSDPs and RHs in *Arabidopsis* and rice has provided new opportunities for the mechanistic examination of their cellular roles during the stress adaptation processes. Further studies should focus on RBP-target RNA interactions and RBP-mediated regulation of RNA metabolism, which is an indispensible step in formulating a more comprehensive picture of the cellular functions of RBPs and posttranscriptional regulation of stress adaptation in plants.

Acknowledgments

This work was supported by grants from the National Research Foundation of Korea (2011-0017357 and from the Next-Generation BioGreen 21 Program (PJ00820303), Rural Development Administration, Republic of Korea.

References

1. Dreyfuss G, Matunis MJ, Pinol-Roma S et al. HnRNP proteins and the biogenesis of mRNA. Annu Rev Biochem 1993; 62:289-321.
2. Simpson GG, Filipowicz W. Splicing of precursors to mRNA in higher plants: mechanism, regulation and sub-nuclear organization of the spliceosomal machinery. Plant Mol Biol 1996; 32:1-41.
3. Floris M, Mahgoub H, Lanet E et al. Post-transcriptional regulation of gene expression in plants during abiotic stress. Int J Mol Sci 2009; 10:3168-3185.
4. Burd CG, Dreyfuss G. Conserved structures and diversity of functions of RNA-binding proteins. Science 1994; 265:615-621.
5. Albà MM, Pagès M. Plant proteins containing the RNA-recognition motif. Trends Plant Sci 1998; 3:15-21.
6. Fukami-Kobayashi K, Tomoda S, Go M. Evolutionary clustering and functional similarity of RNA-binding proteins. FEBS Lett 1993; 335:289-293.
7. Lorkovic ZJ, Barta A. Genomic analysis: RNA recognition motif (RRM) and K homology (KH) domain RNA-binding proteins from the flowering plant Arabidopsis thaliana. Nucleic Acids Res 2002; 30:623-635.
8. Barta A, Kalyna M, Reddy ASN. Implementing a rational and consistent nomenclature for serine/arginine-rich protein splicing factors (SR proteins) in plants. Plant Cell 2010; 22:2926-2929.
9. Goldstein J, Pollitt NS, Inouye M. Major cold shock protein of Escherichia coli. Proc Natl Acad Sci U S A 1990; 87:283-287.
10. Thieringer HA, Jones PG, Inouye M. Cold shock and adaptation. BioEssays 1998; 20:49-57.
11. Yamanaka K, Fang L, Inouye M. The CspA family in Escherichia coli: multiple gene duplication for stress adaptation. Mol Micobiol 1998; 27:247-255.
12. Phadtare S, Alsina J, Inouye M. Cold-shock response and cold-shock proteins. Curr Opin Microbiol 1999; 2:175-180.
13. D'Auria G, Esposito C, Falcigno L et al. Dynamical properties of cold shock protein A from Mycobacterium tuberculosis. Biochem Biophys Res Commun 2010; 402:693-698.
14. Fang W, Leger RJ. RNA binding proteins mediate the ability of a fungus to adapt to the cold. Environ Microbiol 2010; 12:810-820.
15. Wolffe AP. Structural and functional properties of the evolutionarily ancient Y-box family of nucleic acid binding proteins. BioEssays 1994; 16:245-251.
16. Jiang W, Hon Y, Inouye M. CspA, the major cold-shock protein of Escherichia coli, is an RNA chaperone J Biol Chem 1997; 272:196-202.
17. Mihailovich M, Militti C, Gabaldón T et al. Eukaryotic cold shock domain proteins: highly versatile regulators of gene expression. Bioessays 2010; 32:109-118.
18. Jankowsky E. RNA helicases at work: binding and rearranging. Trends Biochem Sci 2010; doi:10.1016/j.tibs.2010.07.008.
19. Jones PG, Mitta M, Kim Y et al. Cold shock induces a major ribosomal-associated protein that unwinds doublestranded RNA in Escherichia coli. Proc Natl Acad Sci U S A 1996; 93:76-80.
20. Tanner NK, Cordin O, Banroques J et al. The Q motif: a newly identified motif in DEAD box helicases may regulate ATP binding and hydrolysis. Mol Cell 2003; 11:127-138.
21. Aubourg S, Kreis M, Lecharny A. The DEAD box RNA helicase family in Arabidopsis thaliana. Nucleic Acids Res 1999; 27:628-636.
22. Kenan DJ, Query CC, Keene JD. RNA recognition: towards identifying determinants of specificity. Trends Biochem Sci 1991; 16:214-220.
23. Nagai K, Oubridge C, Ito N et al. The RNP domain: a sequence specific RNA-binding domain involved in processing and transport of RNA. Trends Biochem Sci 1995; 20:235-240.
24. Mangeon A, Junqueira RM, Sachetto-Martins G. Functional diversity of the plant glycine-rich proteins superfamily. Plant Signal Behav 2010; 5:99-104.
25. Kim JY, Kim WY, Kwak KJ et al. Glycine-rich RNA-binding proteins are functionally conserved in Arabidopsis thaliana and Oryza sativa during cold adaptation process. J Exp Bot 2010; 61:2317-2325.
26. Kim JY, Kim WY, Kwak KJ et al. Zinc finger-containing glycine-rich RNA-binding protein in Oryza sativa has an RNA chaperone activity under cold stress conditions. Plant Cell Environ 2010; 33:759-768.
27. Didier DK, Schiffenbauer J, Woulfe SL et al. Characterization of the cDNA encoding a protein binding to the major histocompatibility complex class II Y box. Proc Natl Acad Sci U S A 1988; 85:7322-7326.
28. Landsman D. Rnp-1, an RNA-binding motif is conserved in the DNA-binding cold shock domain. Nucleic Acids Res 1992; 20:2861-2864.

29. Graumann PL, Marahiel MA. A superfamily of proteins that contain the cold-shock domain. Trends Biochem Sci 1998; 23:286-290.
30. Manival X, Ghisolfi-Nieto L, Joseph G et al. RNA-binding strategies common to cold-shock domain and RNA recognition motif-containing proteins. Nucleic Acids Res 2001; 29:2223-2233.
31. Bycroft M, Hubbard TJ, Proctor M et al. The solution structure of the S1 RNA binding domain: a member of an ancient nucleic acid-binding fold. Cell 1997; 88:235-242.
32. Schindelin H, Marahiel MA, Heinemann U. Universal nucleic acid-binding domain revealed by crystal structure of the B. subtilis major cold-shock protein. Nature 1993; 364:154-168.
33. Kloks CP, Spronk CA, Lasonder E et al. The solution structure and DNA-binding properties of the cold-shock domain of the human Y-box protein YB-1. J Mol Biol 2002; 316:317-326.
34. Goroncy AK, Koshiba S, Tochio N et al. The NMR solution structures of the five constituent cold-shock domains (CSD) of the human UNR (upstream of N-ras) protein. J Struct Funct Genomics 2010; 11:181-188.
35. Kingsley PD, Palis J. GRP2 proteins contain both CCHC zinc fingers and a cold shock domain. Plant Cell 1994; 6:1522-1523.
36. Karlson D, Nakaminami K, Yoyomasu T et al. A cold-regulated nucleic acid-binding protein of winter wheat shares a domain with bacterial cold shock proteins. J Biol Chem 2002; 277:35248-35256.
37. Karlson D, Imai R. Conservation of the cold shock domain protein family in plants. Plant Physiol 2003; 131:12-15.
38. Chaikam V, Karlson D. Comparison of structure, function and regulation of plant cold shock domain proteins to bacterial and animal cold shock domain proteins. BMB Rep 2010; 43:1-8.
39. Chaikam V, Karlson D. Functional characterization of two cold shock domain proteins from Oryza sativa. Plant Cell Environ 2008; 31:995-1006.
40. Gorbalenya AE, Koonin EV. Helicases: amino acids sequence comparisons and structure–function relationship. Curr Opin Struct Biol 1993; 3:419-429.
41. Singleton MR, Dillingham MS, Wigley DB. Structure and mechanism of helicases and nucleic acid translocases. Ann Rev Biochem 2007; 76:23-50.
42. Walker JE, Saraste M, Runswick MJ et al. Distantly related sequences in the alpha- and beta-subunits of ATP synthase, myosin, kinases and other ATP-requiring enzymes and a common nucleotide binding fold. EMBO J 1982; 1:945-951.
43. Tuteja N. Plant cell and viral helicases: essential enzymes for nucleic acid transactions. Crit Rev Plant Sci 2000; 19:449-478.
44. Hall MC, Matson SW. Helicase motifs: the engine that powers DNA unwinding. Mol Microbiol 1999; 34:867-877.
45. Rocak S, Linder P. DEAD-box proteins: the driving forces behind RNA metabolism. Nat Rev Mol Cell Biol 2004; 5:232-241.
46. Pause A, Sonenberg N. Mutational analysis of a DEAD-box RNA helicase; the mammalian translation initiation factor eIF-4A. EMBO J 1992; 11:2643-2654.
47. Luking A, Stahl U, Schmidt U. Protein family of RNA helicases. Crit Rev Biochem Mol Biol 1998; 33:259-296.
48. Fairman-Williams ME, Guenther UP, Jankowsky E. SF1 and SF2 helicases: family matters. Curr Opin Struct Biol 2010; 20:313-324.
49. Caruthers JM, Johnson ER, McKay DB. Crystal structure of yeast initiation factor 4A, a DEAD-box RNA helicase. Proc Natl Acad Sci U S A 2000; 97:13080-13085.
50. Story RM, Li H, Abelson JN. Crystal structure of a DEAD box protein from the hyperthermophile Methanococcus jannaschii. Proc Natl Acad Sci USA 2001; 98:1465-1470.
51. Zhao R, Shen J, Green MR et al. Crystal structure of UAP56, a DExD/H-box protein involved in pre-mRNA splicing and mRNA export. Structure 2004; 12:1373-1381.
52. Cheng Z, Coller J, Parker R et al. Crystal structure and functional analysis of DEAD-box protein Dhh1p. RNA 2005; 11:1258-1270.
53. Boudet N, Aubourg S, Toffano-Nioche C et al. Evolution of intron/exon structure of DEAD helicase family genes in Arabidopsis, Caenorhabditis, and Drosophila. Genome Res 2001; 11:2101-2114.
54. Mingam A, Toffano-Nioche C, Brunaud V et al. DEAD-box RNA helicases in Arabidopsis thaliana:establishing a link between quantitative expression, gene structure and evolution of a family of genes. Plant Biotechnol J 2004; 2:401-415.
55. Umate P, Tuteja R, Tuteja N. Genome-wide analysis of helicase gene family from rice and Arabidopsis: a comparison with yeast and human. Plant Mol Biol 2010; 73:449-465.
56. Gómez J, Sánchez-Martínez D, Stiefel V et al. A gene induced by the plant hormone abscisic acid in response to water stress encodes a glycine-rich protein. Nature 1988; 344:262-264.

57. Hirose T, Sugita M, Sugiura M. Characterization of a cDNA encoding a novel type of RNA-binding protein in tobacco: its expression and nucleic acid-binding properties. Mol Gen Genet 1994; 244:360-366.
58. van Nocker S, Vierstra RD. Two cDNAs from Arabidopsis thaliana encode putative RNA binding proteins containing glycine-rich domains. Plant Mol Biol 1993; 21:695-699.
59. Carpenter CD, Kreps JA, Simon AE. Genes encoding glycine-rich Arabidopsis thaliana proteins with RNA-binding motifs are influenced by cold treatment and an endogenous circadian rhythm. Plant Physiol 1994; 104:1015-1025.
60. Ferullo J-M, Vézina LP, Rail J et al. Differential accumulation of two glycine-rich proteins during cold-acclimation alfalfa. Plant Mol Biol 1997; 33:625-633.
61. Molina A, Mena M, Carbonero P et al. Differential expression of pathogen-responsive genes encoding two types of glycine-rich proteins in barley. Plant Mol Biol 1997; 33:803-810.
62. Moriguchi K, Sugita M, Sugiura M. Structure and subcellular localization of a small RNA-binding protein from tabacco. Plant J 1997; 6:825-834.
63. Horvath DP, Olson PA. Cloning and characterization of cold-regulated glycine-rich RNA-binding protein genes from leafy spurge (Euphorbia esula L.) and comparison to heterologous genomic clones. Plant Mol Biol 1998; 38:531-538.
64. Aneeta NS-M, Tuteja N, Sopory SK. Salinity- and ABA-induced up-regulation and light-mediated modulation of mRNA encoding glycine-rich RNA-binding protein from Sorghum bicolor. Biochem Biophys Res Commun 2002; 296:1063-1068.
65. Stephen JR, Dent KC, Finch-Savage WE. A cDNA encoding a cold-induced glycine-rich RNA binding protein from Prunus avium expressed in embryonic axes. Gene 2003; 320:177-183.
66. Nomata T, Kabeya Y, Sato N. Cloning and characterization of glycine-rich RNA-binding protein cDNAs in the moss Physcomitrella patens. Plant Cell Physiol 2004; 45:48-56.
67. Shinozuka H, Hisano H, Yoneyama S et al. Gene expression and genetic mapping analyses of a perennial ryegrass glycine-rich RNA-binding protein gene suggest a role in cold adaptation. Mol Genet Genom 2006; 275:399-408.
68. Sachetto-Martins G, Franco LO, de Oliveira DE. Plant glycine-rich proteins: a family or just proteins with a common motif? Biochim Biophys Acta 2000; 1492:1-14.
69. Lorkovic ZJ. Role of plant RNA-binding proteins in development, stress response and genome organization. Trends Plant Sci 2009; 14:229-236.
70. Kwak KJ, Kim YO, Kang H. Characterization of transgenic Arabidopsis plants overexpressing GR-RBP4 under high salinity, dehydration, or cold stress. J Exp Bot 2005; 56:3007-3016.
71. Kim JY, Park SJ, Jang B et al. Functional characterization of a glycine-rich RNA-binding protein2 in Arabidopsis thaliana under abiotic stress conditions. Plant J 2007; 50:439-451.
72. Kim JS, Jung HJ, Lee HJ et al. Glycine-rich RNA-binding protein7 affects abiotic stress responses by regulating stomata opening and closing in Arabidopsis thaliana. Plant J 2008; 55:455-466.
73. Masaki S, Yamada T, Hirasawa T et al. Proteomic analysis of RNA-binding proteins in dry seeds of rice after fractionation by ssDNA affinity column chromatography. Biotech Lett 2008; 30:955-960.
74. Kim YO, Kim JS, Kang H. Cold-inducible zinc finger-containing glycine-rich RNA-binding protein contributes to the enhancement of freezing tolerance in Arabidopsis thaliana. Plant J 2005; 42:890-900.
75. Kim YO, Kang H. The role of a zinc finger-containing glycine-rich RNA-binding protein during the cold adaptation process in Arabidopsis thaliana. Plant Cell Physiol 2006; 47:793-798.
76. Kim YO, Pan SO, Jung C-H et al. A zinc finger-containing glycine-rich RNA-binding protein, atRZ-1a, has a negative impact on seed germination and seedling growth of Arabidopsis thaliana under salt or drought stress conditions. Plant Cell Physiol 2007; 48:1170-1181.
77. Kim WY, Kim JY, Jung HJ et al. Comparative analysis of Arabidopsis zinc finger-containing glycine-rich RNA-binding proteins during cold adaptation. Plant Physiol Biochem 2010; 48:866-872.
78. Bae W, Xia B, Inouye M et al. Escherichia coli CspA-family RNA chaperones are transcription antiterminators. Proc Natl Acad Sci U S A 2000; 97:7784-7789.
79. Fusaro AF, Bocca SN, Ramos RLB et al. AtGRP2, a cold-induced nucleo-cytoplasmic RNA-binding protein, has a role in flower and seed development. Planta 2007; 225:1339-1351.
80. Sasaki K, Kim MH, Imai R. Arabidopsis cold shock domain protein2 is a RNA chaperone that is regulated by cold and developmental signals. Biochem Biophys Res Commun 2007; 364:633-638.
81. Nakaminami K, Karlson D, Imai R. Functional conservation of cold shock domains in bacteria and higher plants. Proc Natl Acad Sci U S A 2006; 103:10122-10127.
82. Kim M-H, Sasaki K, Imai R. Cold shock domain protein 3 regulates freezing tolerance in Arabidopsis thaliana. J Biol Chem 2009; 284:23454-23460.
83. Park SJ, Kwak KJ, Oh TR et al. Cold shock domain proteins affect seed germination and growth of Arabidopsis thaliana under abiotic stress conditions. Plant Cell Physiol 2009; 50:869-878.

84. Nakaminami K, Hill K, Perry SE et al. Arabidopsis cold shock domain proteins: relationships to floral and silique development. J Exp Bot 2009; 60:1047-1062.
85. Chamot D, Owttrim GW. Regulation of cold shock-induced RNA helicase gene expression in the Cyanobacterium Anabaena sp. Strain PCC 7120. J Bacteriol 2000; 182:1251-1256.
86. Lim J, Thomas T, Cavicchioli R. Low temperature regulated DEAD-box RNA helicase from the Antarctic archaeon, Methanococcoides burtonii. J Mol Biol 2000; 297:553-567.
87. Kreps JA, Wu Y, Chang HS et al. Transcriptome changes for Arabidopsis in response to salt, osmotic, and cold stress. Plant Physiol 2002; 130:2129-2141.
88. Fowler S, Thomashow MF. Arabidopsis transcriptome profiling indicates that multiple regulatory pathways are activated during cold acclimation in addition to the CBF cold response pathway. Plant Cell 2002; 14:1675-1690.
89. Gong Z, Lee H, Xiong L et al. RNA helicase-like protein as an early regulator of transcription factors for plant chilling and freezing tolerance. Proc Natl Acad Sci USA 2002; 99:11507-11512.
90. Gong Z, Dong C-H, Lee H et al. A dead box RNA helicase is essential for mRNA export and important for development and stress responses in Arabidopsis. Plant Cell 2005; 17:256-267.
91. Zhu J, Dong C-H, Zhu J-K. Interplay between cold responsive gene regulation, metabolism and RNA processing during plant cold acclimation. Curr Opin Plant Biol 2007; 10:290-295.
92. Kant P, Kant S, Gordon M et al. STRESS RESPONSE SUPPRESSOR1 and STRESS RESPONSE SUPPRESSOR2, two DEAD-box RNA helicases that attenuate Arabidopsis responses to multiple abiotic stresses. Plant Physiol 2007; 145:814-830.
93. Kim JS, Kim KA, Oh TR et al. Functional characterization of DEAD-box RNA helicases in Arabidopsis thaliana under abiotic stress conditions. Plant Cell Physiol 2008; 49:1563-1571.
94. Nakamura T, Muramoto Y, Takabe T. Structural and transcriptional characterization of a salt-responsive gene encoding putativeATP-dependent RNA helicase in barley. Plant Sci 2004; 167:63-70.
95. Briolat V, Reysset G. Identification of the Clostridium perfringens genes involved in the adaptive response to oxidative stress. J Bacteriol 2002; 184:2333-2343.
96. Li D, Liu H, Zhang H et al. OsBIRH1, a DEAD-box RNA helicase with functions in modulating defence responses against pathogen infection and oxidative stress. J Exp Bot 2008; 59:2133-2146
97. Montero-Lomeli M, Morais BL, Figueiredo DL et al. The initiation factor eIF4A is involved in the response to lithium stress in Saccharomyces cerevisiae. J Biol Chem 2002; 277:21542-21548.
98. Herschlag D. RNA chaperones and the RNA folding problem. J Biol Chem 1995; 270:20871-20874.
99. Xia B, Ke H, Inouye M. Acquirement of cold sensitivity by quadruple deletion of the cspA family and its suppression by PNPase S1 domain in Escherichia coli. Mol Microbiol 2001; 40:179-188.
100. Graumann PL, Wendrich TM, Weber MH et al. A family of cold shock proteins in Bacillus subtilis is essential for cellular growth and for efficient protein synthesis at optimal and low temperatures. Mol Microbiol 1997; 25:741-756.
101. Phadtare S, Inouye M, Severinov K. The nucleic acid melting activity of Escherichia coli CspE is critical for transcription antitermination and cold acclimation of cells. J Biol Chem 2002; 277:7239-7245.
102. Maruyama K, Sato N, Ohta N. Conservation of structure and cold-regulation of RNA-binding proteis in cyanobacteria: probable convergent evolution with eukaryotic glycine-rich RNA-binding proteins. Nucleic Acid Res 1999; 27:2029-2036.
103. Kim JS, Park SJ, Kwak KJ et al. Cold shock domain proteins and glycine-rich RNA-binding proteins from Arabidopsis thaliana can promote the cold adaptation process in Escherichia coli. Nucleic Acids Res 2007; 35:506-516.
104. Park SJ, Kwak KJ, Jung HJ et al. The C-terminal zinc finger domain of Arabidopsis cold shock domain proteins is important for RNA chaperone activity during cold adaptation. Phytochem 2010; 71:543-547.
105. Tanner NK, Linder P. DExD/H box RNA helicases: from generic motors to specific dissociation functions. Mol Cell 2001; 8:251-262.
106. Lorsch JR. RNA chaperones exist and DEAD-box proteins get a life. Cell 2002; 109:797-800.
107. Hanano S, Sugita M, Sugiura M. Isolation of a novel RNA-binding protein and its association with a large ribonucleoprotein particle present in the nucleoplasm of tobacco cells. Plant Mol Biol 1996; 31:57-68.
108. Ludevid MD, Freire MA, Gómez J et al. RNA binding characteristics of a 16 kDa glycine-rich protein in maize. Plant J 1992; 2:999-1003.
109. Hirose T, Sugita M, Sugiura M. cDNA structure, expression and nucleic-acid binding properties of three RNA-binding proteins in tobacco: occurrence of tissue alternative splicing. Nucleic Acids Res 1993; 21:3981-3987.
110. Freire MA, Pagès M. Functional characterization of the maize RNA-binding protein MA16. Plant Mol Biol 1995; 29:797-807.

111. Dunn MA, Brown K, Lightowlers R et al. A low-temperature-responsive gene from barley encodes a protein with single-stranded nucleic acid-binding activity which is phosphorylated in vitro. Plant Mol Biol 1996; 30:947-959.

112. Vermel M, Guermann B, Delage L et al. A family of RRM-type RNA-binding proteins specific to plant mitochondria. Proc Natl Acad Sci U S A 2002; 99:5866-5871.

113. Streitner C, Hennig L, Korneli C et al. Global transcript profiling of transgenic plants constitutively overexpressing the RNA-binding protein AtGRP7. BMC Plant Biol 2010; 10:221.

114. Albà MM. Culiáñez-Macià FA, Goday A et al. The maize RNA-binding protein MA16 is a nucleolar protein located in the dense fibrillar component. Plant J 1994; 6:825-834.

115. Chinnusamy V, Gong Z, Zhu J-K. Nuclear RNA export and its importance in abiotic stress responses of plants. In: Reddy ASN, Golovkin M, eds. Nuclear pre-mRNA processing in plants, Berlin, Heidelberg: Springer-Verlag 2008:235-255.

CHAPTER 9

From Structure to Function of RNA Binding Domains

Antoine Cléry and Frédéric H.-T. Allain*

Abstract

RNA binding proteins (RBPs) are involved in each step of RNA metabolism. Most of them are composed of small RNA binding domains (RBDs) that are needed for their recruitment to specific RNA targets. The mode of RNA recognition of these RBDs has been studied by structural biologists for more than 20 years and seems to be highly versatile. In this chapter we review the current structural knowledge about RNA recognition by the four main RBD families, namely RNA recognition motifs (RRMs), zinc fingers, KH domains and double-stranded RNA binding motifs (dsRBMs) detailing how structural data have brought information essential for a better understanding of RBP functions.

Introduction

This book focuses on the extreme variety of functions carried by RNA binding proteins (RBPs). The association of RBPs with RNA transcripts begins during transcription. Some of these early-binding RBPs remain bound to the RNA until they are degraded, whereas others recognize and transiently bind to RNA at later stages for specific processes such as splicing, processing, transport and localization.[1] Some RBPs also function as RNA chaperones[2] by helping the RNA, which is initially single-stranded, to form various secondary or tertiary structures. When folded, these structured RNAs, together with specific RNA sequences, act as a signal for other RBPs that mediate gene regulation. Most of RNA binding proteins contain several domains including different types of RNA binding motifs, very often in multiple copies (Fig. 1), which recognize RNA sequence specifically.

Here, we review our current structural understanding of protein–RNA recognition mediated by four RNA-binding domains, the RNA recognition motif (RRM), the zinc-finger domain, the KH domain and the double-stranded RNA-binding motif (dsRBM). We discuss how these four small domains recognize RNA. Some bind single-stranded RNA by direct readout of the primary sequence, whereas others recognize primarily the shape of the RNA or both the sequence and the shape. This chapter shows how, within the past 15 years, structural biology revealed the highly versatile mode of protein-RNA interactions and contributed to explain their importance for RBP functions.

RNA Recognition Motifs (RRMs)

The RNA-recognition motif (RRM), also known as RBD (RNA binding domain) or RNP (ribonucleoprotein domain) is the most abundant RNA-binding domain in higher vertebrates (this motif is present in about 0.5%-1% of human genes)[3] and is the most extensively studied

*Institute for Molecular Biology and Biophysics, ETH Zürich, Zürich, Switzerland.
Corresponding Author: Frédéric H.T. Allain—Email: allain@mol.biol.ethz.ch

RNA Binding Proteins, edited by Zdravko J. Lorković.

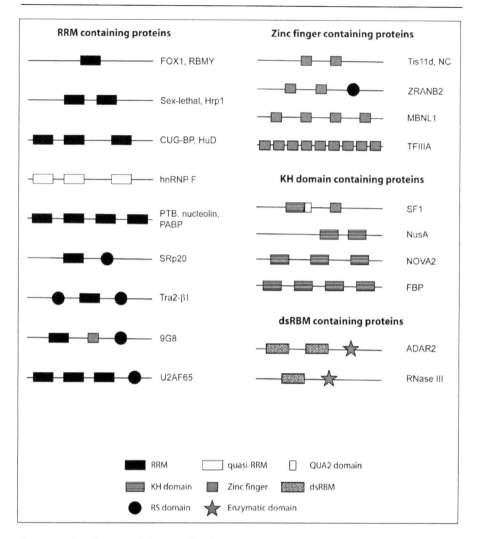

Figure 1. Classification of the RNA binding proteins described in the chapter according to their RNA binding domain composition. RRMs, quasi-RRMs (qRRMs), KH domains, zinc fingers, dsRBMs, enzymatic domains, QUA2 and RS domains are schematically represented as indicated in the figure.

RNA-binding domain, both in terms of structure and biochemistry.[4] Although this domain was also shown interacting with DNA and protein partners, we focus here on its role as an RNA binding domain. Typically, an RRM can be recognized at the primary sequence level as a 90 amino acids long domain containing two conserved sequences of eight and six amino-acids, called RNP1 and RNP2, respectively (Fig. 2A). RRM adopts a typical $\beta_1\alpha_1\beta_2\beta_3\alpha_2\beta_4$ topology that forms a four-stranded β-sheet packed against two α-helices (Fig. 2B). The RNP1 and RNP2 sequences located in the two central β-strands β3 and β1 of the domain, respectively, expose three conserved aromatic residues on the surface of the β-sheet which form the primary RNA binding surface. Lack of the presence of most if not all of these aromatic residues led to the definition of several subclass of RRMs like the quasi-RRM (qRRM), the pseudo-RRM (ΨRRM) or U2AF Homology Motifs (UHM).[5]

Since the first structure of an RRM-RNA complex in 1994 by Nagai and coworkers,[6] more than 30 high-resolution structures of RRM-RNA complexes have been determined either by X-ray crystallography[6-19] or NMR spectroscopy[20-33] (Table 1). This fairly large ensemble of available structures revealed common features between the complexes and also many surprises.

A Simple Fold Binding a Large Panel of RNA Sequences and Structures

Binding the β-Sheet Surface

As maybe expected from sequence conservation, the four-stranded β-sheet constitutes the primary and most common RNA binding surface of an RRM. Typically, the three conserved aromatic side-chains located in the conserved sequences RNP1 (in β3-strand) and RNP2 (in β1-strand) (Fig. 2A), accommodate two nucleotides as follows: the bases of the 5′ and of the 3′ nucleotides stack on an aromatic ring located in β1 (position 2 of RNP2) and in β3 (position 5 of RNP1), respectively (Fig. 2B). The third aromatic ring located in β3 (position 3 of RNP1) is found often inserted between the two sugar rings of the dinucleotide (Fig. 2B). The RRM β-sheet surface is therefore a perfect platform to interact with two consecutive single-stranded RNA nucleotides. Three to four nucleotides are usually accommodated on the β-sheet, via the presence of aromatic rings or other planar side-chains (Arg, Asn, Asp, His) on the other β-strands as shown in the RRM structure of hnRNPA1 (Fig. 2B) or SRp20 (Fig. 2C). If these aromatic rings and planar side-chains provide RNA binding affinity, they do not achieve sequence-specificity. RNA sequence-specificity by RRMs is achieved by side-chains present on the β-sheet surface but also by the main-chain of the residues immediately C-terminal to the RRM (i.e., the few residues following strand β4) as shown in the case of the RRM of hnRNPA1 (Fig. 2B) or SRp20 (Fig. 2C). This mode of binding allows a binding affinity in the micromolar range and a certain bias in sequence-specificity since the involvement of the main-chain in binding results in a binding preference of the canonical RRM for a W-G dinucleotide (W can be A or C).[34] Although most of RRMs use this basic mode of binding, RRM evolved to be capable of binding a large repertoire of sequences by using parts of the domain outside the strict β-sheet surface.

The Extension of the β-Sheet Surface

One way to accommodate more nucleotides and therefore achieve higher affinity is to extend the β-sheet surface of the RRM. To date, two RRMs were shown to have such extended surface, the RRM2 and RRM3 of PTB which both contain a fifth β-strand anti parallel to β2 (Fig. 2D). The presence of this additional strand allowed the binding of one (RRM2) or two (RRM3) additional nucleotides.[28] A second approach selected across evolution was to juxtapose two consecutive RRMs as in PAPB where β2 of RRM2 interacts with β4 of RRM1 creating a continuous surface that can accommodate eight adenines[7] (Fig. 3).

The Critical Role of Protein Loops to Extend RNA Binding

In other RRMs, the loops connecting the β-strands or the β-strands to the α-helices of the RRMs can be involved in RNA recognition. The most spectacular example in this context is probably the RRM of Fox-1 where the β1-α1 and α2-β4 loops of human Fox-1 RRM[22] are involved in the binding of the first four nucleotides of the sequence 5′-UGCAUGU-3′. In particular, one Phenylalanine located in the β1-α1 loop stacks with the first two RNA nucleotides, whereas the last three nucleotides are recognized in a canonical manner by the β-sheet of the RRM (Fig. 2E). This structure nicely shows how nonconserved parts of the RRM can be used to extend to more than double of the canonical binding surface. As a result Fox-1 RRM binds the heptamer 5′-UGCAUGU-3′ with subnanomolar affinity.

In addition to be involved in the specific binding of RNA sequences, RRM loops can also be responsible for the recognition of the RNA shape. This is probably best illustrated by the structure of RBMY RRM in complex with an RNA stem-loop, where all residues of the β2-β3 loop insert into the RNA major groove[30] of the RNA helix while the β-sheet specifically recognizes the RNA loop (Fig. 2F). Similarly, the β2-β3 loop of the N-terminal RRM of U1A or of U2B″ is also crucial to recognize a stem-loop[6,14] or an internal loop.[35]

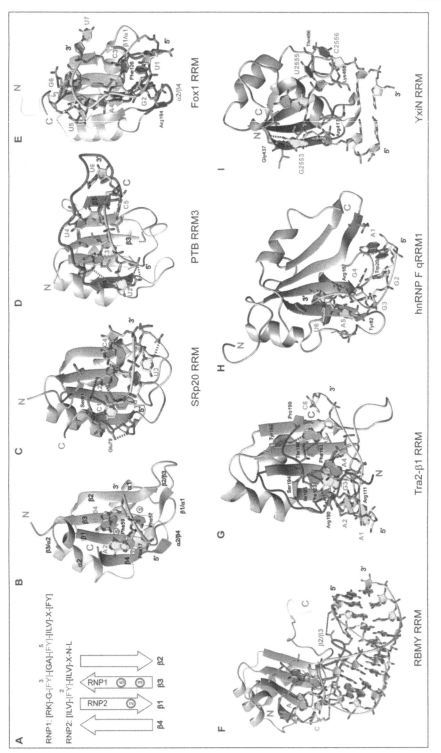

Figure 2. Please see the figure legend on the following page.

Figure 2, viewed on previous page. The high versatility of single RRM interactions with RNA. A) Scheme of the four-stranded β-sheet with the position and sequences of main conserved RNP1 and RNP2 aromatic residues shown in green. RNP1 and RNP2 consensus sequences of RRMs are shown (X is for any amino acid). B) Structure of hnRNP A1 RRM2 in complex with single stranded telomeric DNA as a model of single stranded nucleic acid binding.[89] C) Structure of SRp20 RRM in complex with the 5'-CAUC-3' RNA.[25] D) Structure of PTB RRM3 in complex with the 5'-CUCUCU-3' RNA.[28] E) Structure of Fox-1 RRM in complex with the 5'-UGCAUGU-3' RNA.[22] F) Structure of RBMY RRM in complex with a stem-loop RNA capped by a 5'-CACAA-3' pentaloop.[30] G) Structure of Tra2-β1 RRM in complex with the 5'-AAGAAC-3' RNA.[23] H) Structure of hnRNP F qRRM1 in complex with the 5'-AGGGAU-3' RNA.[24] I) Structure of YxiN RRM in complex with RNA.[9] In all the figures, the ribbon of the RRM is shown in grey, the RNA nucleotides are in yellow and the protein side-chains are in green. The N, O and P atoms are in blue, red and orange, respectively. The N- and C-terminal extensions of the RRM and 5'- and 3'-end of RNA are indicated. Hydrogen bonds are represented by purple dashed lines. All the figures were generated by the program MOLMOL.[90]

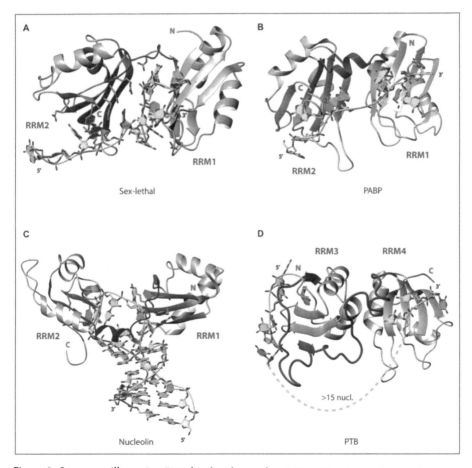

Figure 3. Structures illustrating RNA binding by tandem RRMs. Structures of Sex-lethal[8] (A), PABP[7] (B), Nucleolin[20] (C) and PTB[28] (D) RRMs are shown in complex with RNA. Linker separating RRMs is shown red. Figures were generated by the program MOLMOL[90] using the same colour schemes that in the Figure 2.

Table 1. *Structures of RRMs solved in complex with RNA. The protein domains and the method used for the structure determination are indicated with the corresponding PDB number (http://www.rcsb.org/pdb/home/home.do).*

Structure Title	Exp. Method	PDB ID	Reference
Spliceosomal U2B"-U2A' proteins bound to a fragment of the U2 snRNA	X-RAY DIFFRACTION	1A9N	14
U1A in complex with the RNA polyadenylation inhibition element	SOLUTION NMR	1AUD	21
U1A in complex with an RNA hairpin	X-RAY DIFFRACTION	1URN	6
U1A in complex with PIE RNA	SOLUTION NMR	1DZ5	33
Sex-lethal in complex with the tra mRNA precursor	X-RAY DIFFRACTION	1B7F	8
PABP in complex with a polyA tract RNA	X-RAY DIFFRACTION	1CVJ	7
Nucleolin RRM1 + 2 in complex with the SNRE RNA	SOLUTION NMR	1FJE	20
Nucleolin RRM1 + 2 in complex with a pre-rRNA target	SOLUTION NMR	1RKJ	26
HuC RRM1+2 in complex with a AU-rich element (ARE)	SOLUTION NMR	1FNX	
HuD RRM1 + 2 in complex with a AU-rich element (class I ARE)	X-RAY DIFFRACTION	1FXL	18
HuD RRM1 + 2 in complex with a AU-rich element (class II ARE)	X-RAY DIFFRACTION	1G2E	18
PTB RRM1 in complex with the CUCUCU RNA	SOLUTION NMR	2AD9	28
PTB RRM2 in complex with the CUCUCU RNA	SOLUTION NMR	2ADB	28
PTB RRM3 + 4 in complex with the CUCUCU RNA	SOLUTION NMR	2ADC	28
Hrp1 RRM in complex with the UAUAUAUA RNA	SOLUTION NMR	2CJK	29
Fox-1 RRM in complex with the UGCAUGU RNA	SOLUTION NMR	2ERR	22
RBMY RRM in complex with a RNA stem loop	SOLUTION NMR	2FY1	30
U2AF65 RRM in complex with a polyuridine tract	X-RAY DIFFRACTION	2G4B	15
SRp20 RRM in complex the CAUC RNA	SOLUTION NMR	2I2Y	25

continued on next page

Table 1. Continued

Structure Title	Exp. Method	PDB ID	Reference
hnRNP F qRRM1 in complex with a G-tract RNA	SOLUTION NMR	2KFY	24
hnRNP F qRRM2 in complex with a G-tract RNA	SOLUTION NMR	2KG0	24
hnRNP F qRRM3 in complex with a G-tract RNA	SOLUTION NMR	2KG1	24
Prp24 RRM2 bound to a fragment of the U6 snRNA	SOLUTION NMR	2KH9	27
Tra2beta1 RRM in complex with the AAGAAC RNA	SOLUTION NMR	2KXN	23
Tra2beta1 RRM in complex with the GAAGAA RNA	SOLUTION NMR	2RRA	32
Nab3 RRM in complex with the UCUU RNA	SOLUTION NMR	2L41	
La in complex with the UGCUGUUUU RNA	X-RAY DIFFRACTION	1ZH5	17
La in complex with the AUAUUU RNA	X-RAY DIFFRACTION	2VOD	10
La in complex with the AUAAUU RNA	X-RAY DIFFRACTION	2VON	10
La in complex with the UUUUUUU RNA	X-RAY DIFFRACTION	2VOO	10
La in complex with the AUUUU RNA	X-RAY DIFFRACTION	2VOP	10
RNA15 RRM in complex with the GUUGU RNA	X-RAY DIFFRACTION	2X1A	12
Nab3 RRM bound to the UUCUUAUUCUUA RNA	X-RAY DIFFRACTION	2XNR	11
CUGBP1 RRM1 in complex with the GUUGUUUUGUUU RNA	X-RAY DIFFRACTION	3NNH	16
CUGBP1 RRM1 + 2 in complex with the UGUGUGUUGUGUG RNA	X-RAY DIFFRACTION	3NNC	16
CUGBP1 RRM3 in complex with the UGUGUG RNA	SOLUTION NMR	2RQC	31
DEAD box helicase YxiN in complex with a 23S rRNA fragment	X-RAY DIFFRACTION	3MOJ	9
CFIm68 RRM/CFIm25 in complex with RNA	X-RAY DIFFRACTION	3Q2T	19
Human Spliceosomal U1 snRNP	X-RAY DIFFRACTION	3CW1	13

The Use of RRM N- and C-Terminal Regions

The N- and C-terminal regions, outside the RRM, are usually poorly ordered in the isolated domains with a few exceptions where they can adopt a secondary structure. For example, in the structures of the La C-terminal RRM,[36] the N-terminal RRM of U1A[37] of CstF-64 RRM[38] or more recently in the qRRM of hnRNP F,[24] the C-terminus forms a α-helix that covers the β-sheet surface of the RRM preventing its access to RNA. However, in an increasing number of RRMs, the regions outside the predicted RRMs have proved to be of crucial importance to significantly enhance RNA binding affinity and to play a role in sequence-specificity. Several structures show that the N-terminal (CUG-BP1, CBP20),[31,39] the C-terminal (PABP and PTB)[7,28] or both the N- and C-termini (Tra2-β1)[23,32] regions of the RRM can be used to directly interact with the targeted RNA. In this latter example, both N- and C-terminal regions of Tra2-β1 RRM are found unstructured in the free RRM and become ordered only upon RNA binding. Both terminal regions interact with RNA and cross each other[23,32] (Fig. 2G). Amino acids from these extremities and from the β-sheet participate to the specific recognition of the 5'-AGAA-3' sequence (Fig. 2G). This positioning of two termini induced upon RNA binding in Tra2-β1 could be functionally important as it could explain how Tra2-β1 recruits two additional proteins, hnRNPG and SRp30c, on SMN (survival of motor neuron) exon 7[23] in order to increase its splicing.[40,41]

The qRRM, an RRM Binding RNA without Using the β-Sheet Surface

In contrast to the RRMs discussed above which always use the canonical β-sheet surface, the qRRMs of hnRNP F bind RNA in a different way. Indeed, the β1/α1, β2/β3 and α2/β4 loops, but not the β-sheet, interact with RNA.[24] Due to the lack of conserved aromatic residues in RNP1 and RNP2, the three RRMs of hnRNP F have been renamed qRRMs for quasi RRMs.[42] The recent structures of the three RRMs of hnRNP F in complex with a G-tract of three guanines show that each qRRM binds RNA in an identical manner. The three guanines adopt a compact conformation surrounded by three conserved residues belonging to protein loops that are stacking with each guanine base (two aromatics and one arginine). The qRRMs appear to encage the G-tract[24] (Fig. 2H). This mode of binding appears conserved among qRRMs, since the side-chains involved in sequence-specific recognition of the G-tract are conserved among identified qRRMs.[24] By sequestering in this manner G triplets, it was proposed that hnRNP F could regulate splicing by maintaining G-rich sequences in a single stranded conformation, therefore preventing RNA to fold into a secondary structure.[24]

The RRM, an Extraordinary Plastic RNA Binding Module

Altogether, these structural investigations showed that the interaction of RRM with RNA is not restricted to the β-sheet surface and that the loops (especially loops 1, 3 and 5) and the two termini appear to be equally important to the β-sheet for binding RNA with high affinity and sequence-specificity. Although, the canonical β-sheet surface appears to interact preferably with a C/A-G sequence, RRMs are found binding almost any types of sequences and secondary structures showing the extraordinary plasticity of this RNA binding domain. In this context, the recent structure of an RRM of the bacterial DEAD-box helicase YxiN bound to RNA perfectly illustrates the extraordinary plasticity of an RRM, since this RRM in order to recognize a three-way RNA junction requires the involvement of almost all parts of the domain: β-sheet, loops and even the helices (Fig. 2I).[9]

When Multiple RRMs Bind RNA

Proteins containing multiple RRMs seem to be the rule more than the exception. Often the RRMs within one protein show high similarity in primary sequence and therefore the RNA binding specificity of each RRM can be almost identical (the three RRMs of hnRNP F or the two RRMs of PABP or hnRNP A1) or similar (the four RRMs of PTB or the three RRMs of CUG-BP). Two RRMs within one protein can be also highly dissimilar in sequence and function as shown for the

RRMs of U1A, U2B″, U2AF65 and U2AF35 or of several SR proteins.[4] Yet, the ways multiple RRMs are used by each proteins to bind RNA seem to differ drastically.

Multiple RRMs to Achieve Higher RNA Binding Affinity and Sequence-Specificity

As one would expect, similarly to what was shown for DNA binding modules, having RRMs in tandem allows higher RNA binding affinity and sequence-specificity compared to single RRM if both domains contribute to RNA binding. This has been nicely illustrated with the structures of several tandem RRMs bound to RNA, namely Sex-lethal,[8] PABP,[7] HuD,[18] nucleolin,[20] or more recently Hrp1.[29] In all five complexes, the two RRMs bind synergistically a continuous single-stranded RNA sequence or stem-loop. The β-sheet of both RRMs is the primary binding surface, however the short (around ten amino-acids) interdomain linker between the two RRMs plays a key role for the RNA recognition. In all five proteins, the interdomain linker is flexible in the free state to become well ordered upon RNA binding. This further illustrates the importance of the N- and C-terminal extension of RRMs for RNA recognition. In all complexes, RRM2 and RRM1 bind the upstream and downstream RNA sequence, respectively and for the complexes of Hrp1,[29] HuD[18] and Sex-lethal[8] the orientation of the two RRMs is almost identical, the two RRM forming a cleft (Fig. 3A). In PABP,[7] the β-sheets of both domains are adjacent forming more a large RNA binding platform (Fig. 3B), while in nucleolin,[20] two RRMs sandwiched the RNA (Fig. 3C).

Multiple RRMs to Affect the RNA Topology

In PTB or in hnRNPA1, tandem RRMs interact already in the free state positioning these consecutive domains in such a way that they cannot bind adjacent sequences. The structures of the complexes with single-stranded RNA and DNA for PTB RRM34 and hnRNPA1 RRM12, respectively, indeed showed that such orientations of the RRMs could induce the formation of RNA loops between the two binding sites of each RRM (Fig. 3D). This mode of binding could explain the mode of action of both proteins as repressor of splicing.[28,43]

When Multiple RRMs Bind Independently, What Is the Function?

In many proteins containing multiple RRMs, the RRMs do not interact with each other and appear to bind RNA independently from each other. In PTB for example, RRM1 and RRM2 are clearly independent from the other RRMs and are separated by long flexible linkers. The same is true for hnRNP F qRRMs (Dominguez and Allain, unpublished). Within each hnRNP protein, each RRM binds very similar sequences (UCU for PTB RRMs and GGG for hnRNP F qRRMs) with micromolar affinity. Part of the function for this independent binding of each RRM might be to increase the chance for the protein to encounter their specific binding sequence. Having long flexible linkers between the RRMs allow the protein to span large volumes and therefore increase the probability to find an accessible RNA binding sequence. The same might hold true for two other splicing factors U2AF65 and CUG-BP for which each RRM appears to bind independently U-rich and GU-rich sequences, respectively.[15,16,31]

When Multi-RRM Proteins Compete Functionally

In light of the three mode of RNA binding described above, it is interesting to compare Sex-lethal, PTB and U2AF both structurally and functionally since these three multi-RRMs bind all the 3′ splice-site pyrimidine-tract yet with a different mode of binding. The fact that all three proteins can function at the same site when the 3′ splice-site sequence is optimal for the binding for each protein[44] although their RNA binding mode is very different argues that a precise binding place dictates the function more than the binding mode. If Sex-lethal has the advantage of binding with higher affinity due to the cooperativity of binding between the two RRMs, U2AF and PTB although they bind more weakly, have the advantage over Sex-lethal to be less stringent with the RNA target they can bind both in term of sequence and length. Similar competitions between RNA binding proteins using a different mode of binding is likely to be frequent in RNA biology which render the molecular mechanisms of posttranscriptional gene regulation difficult to decipher or to model.

The Zinc Finger Domain

Zinc finger (ZnF) is another domain found in RNA binding proteins. In a single RBP this motif can be found alone, as a repeated domain or even in combination with other types of RBDs (Fig. 1). A classical zinc finger is about 30 amino acids long and displays a ββα protein fold in which a β-hairpin and an α-helix are pinned together by a Zn^{2+} ion. These domains are classified depending on the amino acids that interact with this ion (e.g., CCHH, CCCH or CCCC) and were initially described as DNA binding motifs, the CCHH-type being the most frequent one. They were found interacting specifically with dsDNA bases located in major grooves via side chains of residues present in their α-helix.[45] More recently, zinc fingers have also been shown binding RNA molecules. It therefore raised a series of questions. How such a small domain accommodates RNA? Is it able to recognize specifically a RNA sequence? How versatile is this interaction? In this section, referring to the few structures of zinc fingers solved in complex with RNA[46-54] (Table 2), we give a short overview of what is known about the RNA binding mode of these surprising RNA binding motifs.

Table 2. Structures of zinc fingers solved in complex with RNA. The protein domains and the method used for the structure determination are indicated with the corresponding PDB number (http://www.rcsb.org/pdb/home/home.do).

Structure Title	Exp. Method	PDB ID	Reference
HIV-1 nucleocapsid protein bound to the stem-loop sl2	SOLUTION NMR	1F6U	46
HIV-1 nucleocapsid protein bound to the stem-loop sl3	SOLUTION NMR	1A1T	47
MLV nucleocapsid protein bound to the AACAGU RNA	SOLUTION NMR	1WWD	48
MLV nucleocapsid protein bound to the UUUUGCU RNA	SOLUTION NMR	1WWE	48
MLV nucleocapsid protein bound to the CCUCCGU RNA	SOLUTION NMR	1WWF	48
MLV nucleocapsid protein bound to the UAUCUG RNA	SOLUTION NMR	1WWG	48
Rous Sarcoma Virus nucleocapsid Protein bound to RNA	SOLUTION NMR	2IHX	49
TIS11d ZnF1 + 2 in complex with the UUAUUUAUU RNA	SOLUTION NMR	1RGO	50
MBNL1 ZnF3 + 4 in complex with the CGCUGU RNA	SOLUTION NMR	3D2S	54
ZRANB2 ZnF2 in complex with the AGGUAA RNA	SOLUTION NMR	3G9Y	52
TFIIIA ZnF4-6 bound to a minimal 5S RNA	X-RAY DIFFRACTION	1UN6	53
TFIIIA ZnF4-6 bound to a 5S rRNA 55mer	SOLUTION NMR	2HGH	51

The High Diversity of Interaction of Zinc Fingers with RNA

The structure of TFIIIA zinc fingers was among the first to be solved in complex with RNA. TFIIIA is a transcription factor that was shown to be involved in the transcription of eukaryotic ribosomal 5S RNA.[55] This protein contains nine CCHH zinc fingers and can either bind DNA or RNA molecules.[56] The crystal structure of TFIIIA ZnF4 to 6 in complex with a minimal folded version of the 5S rRNA[53] revealed that zinc fingers were able to interact with the backbone of a RNA double-helix and could even specifically recognize individual accessible bases. However, at that time, it was not clear whether these domains could drive proteins to specific ssRNA sequences, as RRMs do. Such evidence came later from other types of zinc fingers (CCCH, CCHC and CCCC) for which the structure was primarily solved in complex with single stranded RNAs (Fig. 4A-E).

Tis11d, a member of the tristetrapolin (TTP) protein family, contains two CCCH zinc fingers (Fig. 1). This protein is involved in the control of the inflammatory response and induces the degradation of mRNAs that contain an AU-rich element (ARE) in their 3′-UTR.[57] In order to understand how this recognition occurs, the structure of the two Tis11d zinc fingers was solved by NMR in complex with the RNA sequence 5′-UUAUUUAUU-3′.[50] This structural study shows that an UAU motif is specifically recognized by each domain through hydrogen bonds mediated primarily by protein backbone atoms to Watson-Crick edges of the bases (Fig. 4B). A similar mode of interaction is used by CCHC zinc fingers of HIV-1 nucleocapsid protein (NC) to recognize a different motif.[47] As illustrated in the Figure 4C, the ZnF1 of NC interacts specifically with a 5′-AG-3′ dinucleotide. These structures demonstrated the ability of zinc fingers to recognize different RNA sequences. However, this RNA recognition mode using almost exclusively the backbone of the protein is not the only way for zinc fingers to specifically bind RNA. The next two examples reveal that these domains are in fact able to bind different RNA sequences with an unexpected diversity of interactions.

MBNL1 (Muscleblind-like 1) contains four CCCH zinc fingers (Fig. 1) and was also shown to bind RNA. This tissue-specific alternative splicing regulator was proposed to promote muscle differentiation.[58] Understanding the mode of action of this protein is a current challenge since its inactivation is in part responsible for leading to the myotonic dystrophy disease. A crystal structure was solved showing that 5′-GC-3′ and 5′-GCU-3′ RNA motifs are specifically recognized by MBNL1 ZnF3 and 4, respectively, with a nanomolar affinity. As for most of the proteins that contain several RNA binding domains, the mode of RNA recognition is very similar for these two zinc fingers. As illustrated in Figure 4D with the ZnF3, this sequence-specific recognition is mediated by stacking interactions and several hydrogen bonds involving main chains and, contrary to Tis11d, also side chains of the protein. A particularity of the MBNL1-RNA complex is that two cysteines, (Cys185 and Cys200) co-ordinating zinc atom, also interact with RNA (Fig. 4D).

The structures of Tis11d and MBNL1 illustrate how two CCCH ZnF containing proteins can specifically recognize different RNA sequences, namely UAU or GC(U) using two distinct modes of interaction. However, the opposite is also observed since several proteins sharing sequence homologies in their zinc finger domains were found interacting with the same RNA sequence. An example is presented below with the atypical ZRANB2 zinc finger family.

ZRANB2 is a human splicing factor that contains two RanBP2-type ("CCCC") zinc finger domains followed by a RS domain (Fig. 1). As described in Chapter 2 by Mueller and Hertel, the RS domains are characterized by a repetition of Arg-Ser dipeptides and are mainly involved in protein-protein interactions. The RanBP2-type domain binds RNA and is defined by the consensus sequence $W-X-C-X_{2-4}-C-X_3-N-X_6-C-X_2-C$. This domain adopts an unusual fold that comprises two short β-hairpins sandwiching a central tryptophan residue and a single zinc ion co-ordinated by four conserved cysteines[59] (Fig. 4E). A crystal structure of the two ZRANB2 ZnFs in complex with the RNA sequence 5′-AGGUAA-3′ was determined recently.[52] A structural particularity of this RNA-protein complex is the guanine-Trp79-guanine "ladder" formation adopted by a continuous stacking of these three residues (Fig. 4E). The three consecutive bases

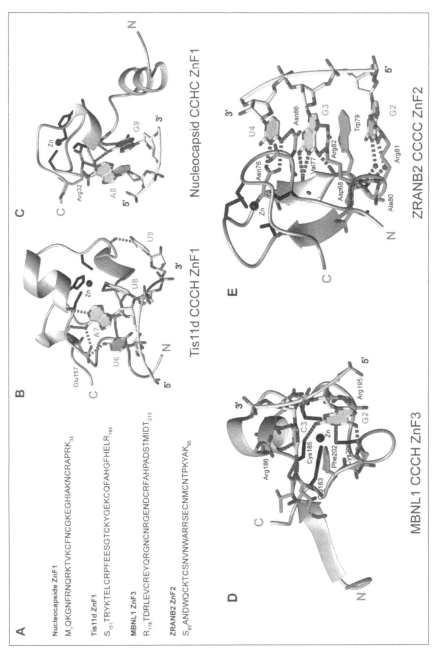

Figure 4A-E. F-H viewed on following page. Please see figure legend on following page.

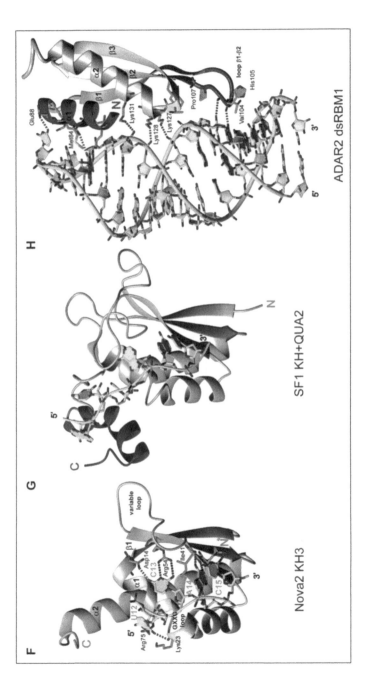

F — Nova2 KH3

G — SF1 KH+QUA2

H — ADAR2 dsRBM1

Figure 4, continued from previous page. Structures of zinc finger, KH domain and dsRBM in complex with RNA. A) Amino-acid sequences of zinc fingers presented in panels B to E. The residues that interact with the zinc ion are in red. B) Solution structure of Tis11d ZnF1 in complex with the 5'-UAUU-3' RNA.[50] C) Solution structure of the ZnF1 of HIV-1 nucleocapsid protein bound to the 5'-AG-3' dinucleotide.[47] D) Crystal structure of MBNL1 ZnF3 bound to the 5'-GC-3' dinucleotide.[54] E) Crystal structure of ZRANB2 ZnF2 in complex with the 5'-GGU-3' RNA.[52] The zinc atom and water molecules are represented by black and red spheres, respectively. Side chains of residues interacting with the zinc ion are in black. F) Crystal structure of Nova2 KH3 bound to the 5'-UCAC-3' RNA.[67] G) Solution structure of SF1 KH and QUA2 domains in complex with the 5'-ACUAAC-3' RNA.[68] The QUA2 domain is represented in red. H) Solution structure of ADAR2 dsRBM1 in complex with a stem-loop RNA.[82] Helix α1 and loop β1-β2 that interact with minor groves are shown in red. Figures were generated by the program MOLMOL[90] using the same colour schemes that in the Figure 2.

G2, G3 and U4 are specifically recognized by formation of hydrogen bonds involving protein side chains (Asn76, Arg81, Arg82 and Asn86), backbone groups (Val77 carbonyl and Trp79 amide) and water-mediated hydrogen bonds (Asp68 and Ala80) (Fig. 4E). These amino acids are mainly located in ZRANB2 loops, especially the one located at the C-terminal extremity of the first β-hairpin (Fig. 4E). Based on these structural data and biochemical studies, six other ZRANB2 ZnF containing proteins involved in mRNA processing could be identified in humans as binding a GGU motif in ssRNAs with micromolar affinities.[60]

Which Functions for Zinc Fingers?

Structural data did not only help to better understand the specificity of interaction of these RNA binding domains and identify new putative binding motifs for ZnF binding proteins, they also helped to find new possible functions for these factors. For ZRANB2, based on functional data and the strong homology between its binding sequence evidenced by the structure and the 5' splice-site sequences, the authors suggested that this protein might interact with a subset of 5' splice-sites preventing their recognition by the spliceosome.[52] Also based on structural data, the anti-parallel orientation observed for the RNA molecules bound by the two zinc fingers of MBNL1 and the location of its binding sites on natural targets suggested that this protein could induce a looping of the RNA, as observed for PTB RRM34, blocking the 3' splice-site recognition by U2 snRNP and resulting in exon skipping.[54]

In this section, we showed that as described for RRMs, zinc fingers are able to specifically bind single-stranded RNA with affinities ranging from nanomolar to the micromolar, using primarily hydrogen bond and aromatic-base stacking interactions. However, the amino acids involved in RNA interaction are not primarily located in the β-strands like for most of RRMs but rather embedded in the protein loops and α-helices. Another characteristic of these domains is their ability to adopt different folds in order to be able to recognize different sequences using their main chains. Due to the apparent wide diversity of RNA-ZnF interactions, more structures are now needed to better understand and classify the different modes of RNA recognition of these small RNA binding domains.

The KH Domain

The hnRNP K homology (KH) domain is approximately 70 amino acids long. It is found in proteins with different functions including splicing, transcriptional regulation and translational control. Two versions of the KH fold have been reported, the Types I and II found in eukaryotic and prokaryotic proteins, respectively. The Type I has a βααββα topology and is characterized by a β-sheet composed of three antiparallel β-strands packed against three α-helices.[61,62] The β1- and β2-strands are parallel to each other and the β3-strand is antiparallel to both. In addition, a "GXXG loop" containing the (I/L/V)-I-G-X-X-G-X-X-(I/L/V) conserved motif, located between the α1 and α2 helices, and a β2-β3 loop variable in length (3 to over 60 amino acids) and sequence, are also found in this motif (Fig. 4F). The KH Type II fold differs from the Type I by a αββααβ topology and a characteristic β-sheet in which the central strand (β2) is parallel to β3 and antiparallel to β1.[61,62] Although both KH motif folds are known for interacting with RNA or ssDNA targets, only few structures of these domains bound to nucleic acid molecules have been deposited in the Protein Data Bank[63-69] (Table 3) and most of them concern the eukaryotic Type I KH domain. Therefore, we will essentially focus on this type of KH fold in this section.

KH Domains Bind Four Nucleotides

KH domains have been shown interacting with their nucleic acid targets using common features. Typically, the single-stranded RNA or DNA molecule is mostly bound by an extended RNA binding surface including the α1 and α2 helices linked by the GXXG loop on one side and the β-sheet and the variable loop on the other side.[62] Together, they form a binding cleft that usually accommodates four bases (Fig. 4F). As an example, we describe the mode of interaction of the KH3 domain of Nova2 (Neuro-oncological ventral antigen 2) with RNA. This tissue-specific alternative splicing factor is highly expressed in the neocortex and hippocampus where it regulates the alternative splicing of

**Table 3. Structures of KH domains solved in complex with RNA. The protein domains
and the method used for the structure determination are indicated with the
corresponding PDB number (http://www.rcsb.org/pdb/home/home.do).**

Structure Title	Exp. Method	PDB ID	Reference
Nova-1 KH1 + 2 bound to a RNA hairpin	X-RAY DIFFRACTION	2ANR	
Nova-2 KH3 bound to a RNA hairpin	X-RAY DIFFRACTION	1EC6	67
SF1 KH and QUA2 domains bound to the UAUACUAACAA RNA	SOLUTION NMR	1K1G	68
PCBP2 KH1 bound to RNA	X-RAY DIFFRACTION	2PY9	65
a/eIF2alpha + aDim2p in complex with the GGAUCACCUCC RNA	X-RAY DIFFRACTION	3AEV	66
ERA KH in complex with RNA	X-RAY DIFFRACTION	3IEV	69
FBP KH3 + 4 in complex with a 29mer ssDNA (derived from FUSE)	SOLUTION NMR	1J4W	64
NusA KH1 + 2 in complex with the GAACUCAAUAG RNA	X-RAY DIFFRACTION	2ASB	63

transcripts coding for proteins having specific functions in brain.[70] Nova2 contains three KH domains of Type I (Fig. 1). The crystal structure of the KH3 domain in complex with an in vitro selected stem-loop RNA shows that this protein interacts with the single stranded 5'-UCAC-3' sequence located in the loop[67] (Fig. 4F). U12 is specifically and indirectly recognized by two water molecules forming hydrogen bonds with the Lys23 and the Arg75 located in the GKGG protein loop and the α3 helix, respectively (Fig. 4F). C13 and C15 directly interact with protein side chains from the β2 and β3 strands, whereas A14 is the only base to be hydrogen bonded to amide and carbonyl of the protein main chain (Ile41) (Fig. 4F). This structure revealed that the NOVA2 KH3 domain interacts specifically with the 5'-UCAY-3' RNA sequence. This information has been crucial for the in vivo identification of many new Nova binding sites and for a better understanding of the splicing regulation by this protein.[70-72]

Since only four bases can be accommodated by KH domains, we wondered how diverse were the sequences recognized by these RBDs. The available structures (Table 3) reveal that the motifs UCAC, UAAC, TCCC, CCCT, AGAA, CAAU, ATTC and TTTT were found interacting with NOVA KH3, SF1 KH, hnRNPK KH3, PCBP2 KH3, NusA KH1 and KH2, FBP KH3 and KH4, respectively. Surprisingly, although the mode of interaction of these domains with nucleic acids seems not to be as versatile as RRMs or ZnFs, KH domains can nevertheless bind a large panel of sequences. Comparing the RNA binding interactions reported for these three RNA binding domains, another difference is the absence of inter RNA-protein stacking interactions described for KH domains. This particularity could in part explain the low affinity (micromolar range) observed for these motifs interacting with single stranded nucleic acids. In order to counteract this apparent low specificity and affinity at least two strategies were selected during the evolution of these RBDs.

How KH Domains Increase Their Affinity and Specificity of Interaction

The first strategy consists in extending the KH domain surface of interaction with nucleic acids. The splicing factor SF1/mBBP is a good example since it specifically binds the 5'-UACUAAC-3' intron branchpoint sequence (BPS) in human pre-mRNA transcripts[73] using a binding surface composed of a KH domain extended by a C-terminal helix known as the QUA2 domain (Quaking homology 2)[68] (Fig. 1). This extended KH surface with a βααββαα topology enables the binding of six nucleotides instead of the four nucleotides usually bound by a single KH domain. The 3'-end of the BPS (5'-UA<u>A</u>C-3'), which contains the conserved branch point adenosine (underlined), is specifically recognized by the KH domain, whereas the 5'-end (5'-ACU-3') is bound by conserved residues from the QUA2 domain (Fig. 4G). In good agreement with the conservation of the branch point adenosine, the NMR structure of this complex shows that this base is specifically recognized by hydrogen bonds involving the main chain of Ile177[68] similar to the contact to A14 in Nova2 KH3 with Ile41 (Fig. 4F). Another example of extension was also reported for the KH4 domain of the KSRP protein which contains a fourth β-strand located adjacent to the β1-strand (2HH2).[74] However, this additional secondary element has still not been shown to be involved in binding nucleic acids.

The second feature consists in the repetition of multiple KH domains within a single RNA binding protein. As described for RRMs, these domains can either act independently or cooperatively. FUSE-binding protein (FBP) contains four KH domains and regulates c-myc expression by binding to FUSE.[75] The structure of FBP KH domains 3 and 4 was solved by NMR in complex with a 29 nucleotide ssDNA molecule derived from its FUSE binding site. Each KH domain binds separately its DNA target. They behave independently without any contact between each other due to the presence of a glycine-rich flexible linker (30 amino acids) separating both KH domains. As observed for RRMs the presence of multiple RBDs in a single protein increases their chance to bind their targets especially when each domain binds nucleic acid molecules with a weak affinity. On the contrary, the two KH domains of NusA are separated by a short linker. It results in extensive contacts between the two domains forming a continuous platform of interaction for the targeted RNA. The 5'-end and 3'-end of the RNA interact within the cleft of KH1 and KH2, respectively binding together RNA with a nanomolar affinity.[63] Similarly, the KH3 and KH4 domains of KSRP were also shown binding their ligand more tightly than each separated.[74]

In this section, we showed that KH domains are able to interact with a large panel of four-nucleotide long sequences. However, contrary to zinc fingers, they always keep the same fold and adapt their recognition mode by subtle variations of the atom sets involved in nucleic acid binding. Importantly, although a single KH domain binds nucleic acid molecules with a rather weak affinity, when present in multiple copies they can act in synergy and interact efficiently with their targets.

The dsRBMs

dsRBMs Bind Double Stranded RNAs

Contrary to the three families of RNA binding domains described above, dsRBMs (double-stranded RNA binding motifs) were first described as recognizing RNA shape rather than RNA sequence.[76] Typically, these domains contain approximately 70 amino acids and exhibit a conserved αββßα protein topology. They are often found in multiple copies (up to five in *Drosophila melanogaster* Staufen protein) and are involved in multiple functions such as RNP localization, RNA interference, RNA processing, RNA localisation, RNA editing and translational control.[77] Until now, only few structures of dsRBMs in complex with RNA have been solved[78-83] (Table 4).

Based on these structures some common features can be observed about the mode of dsRBM interaction with RNA. These domains all interact along one face of a regular A-form helix structure and can cover up to 16 bp (e.g., Xlrbpa2[81]) spanning two consecutive minor grooves separated by a major groove (Fig. 4H). In most of the cases, dsRBMs use residues from the α1 helix and β1-β2 loop (loop 2) to contact the minor grooves and N-terminus of the α2 helix with the preceding loop (loop 4) to bind the major groove (Fig. 4H). In addition, it was previously reported that the spacing between the loops 2 and 4 of dsRBMs fits better with the distance separating the minor and major grooves of RNA A-type helices than with the equivalent distance found in the B-helix form of dsDNA molecules.[77] Based on this last observation and on the multiple interactions described between dsRBMs and 2'-OH groups of RNA riboses, these binding domains were first described as recognizing preferentially the double-stranded RNA shape.[76,77]

Table 4. Structures of dsRBMs solved in complex with RNA. *The protein domains and the method used for the structure determination are indicated with the corresponding PDB number (http://www.rcsb.org/pdb/home/home.do).*

Structure Title	Exp. Method	PDB ID	Reference
Xlrbpa2 bound to RNA	X-RAY DIFFRACTION	1DI2	81
Staufen KH3 bound to RNA	SOLUTION NMR	1EKZ	80
Rnt1p dsRBM in complex with RNA	SOLUTION NMR	1T4O	79
RNase III in complex with RNA	X-RAY DIFFRACTION	2EZ6	78
ADAR2 dsRBM1 in complex with RNA	SOLUTION NMR	2L3C	82
ADAR2 dsRBM2 in complex with RNA	SOLUTION NMR	2L2K	82
TRBP2 dsRBM2 in complex with RNA	X-RAY DIFFRACTION	3ADL	83
HYL1 dsRBM1 in complex with RNA	X-RAY DIFFRACTION	3ADI	83

Interestingly, some particularities have also been emphasized with structures of dsRBMs in complex with RNA. In dsRBM of Xlrbpa2, the α1 helix interacts non specifically with the minor groove of the RNA via few contacts to the bases.[81] These interactions were reported to be mostly mediated by water molecules. In dsRBM3 of Staufen, the α1 helix interacts with a UUCG tetraloop that caps the RNA double helix.[80] In the dsRBM of Rnt1p, the α3 helix stabilizes the conformation of the α1 helix which contacts the sugar-phosphate backbone of the RNA minor groove and two nonconserved bases of the AGNN tetraloop.[84] Finally, the structure of ADAR2 dsRBMs in complex with RNA has recently revealed that these domains were sometimes also able to bind sequence specifically dsRNAs[82] (see below).

Some dsRBMs Interact Specifically with dsRNAs

ADAR2 is a human dsRBM containing protein that converts adenosine-to-inosine (A-to-I) by hydrolytic deamination in numerous mRNA and pre-mRNA transcripts (see Chapter 7 by Jantsch and Vesely).[85,86] This protein has a modular domain organization consisting of two dsRBMs followed by a conserved C-terminal catalytic adenosine deaminase domain (Fig. 1). The structure of ADAR2 dsRBM1 and dsRBM2 was solved by NMR in complex with a stem-loop containing an A-to-I editing site.[82] Both dsRBMs interact similarly with their targeted dsRNA. Lysine residues of a well conserved KKNAK motif located in the N-terminal part of the α2 helix interact nonspecifically with phosphate oxygens of residues from the major groove (Fig. 4H). More unexpected was the fact that each dsRBM binds the RNA stem-loop at a single register with sequence-specific contacts in the minor grooves. The amino group of a guanine is specifically recognized via a hydrogen bond formed with a main chain carbonyl of the β1-β2 loop and a hydrophobic contact is observed between a methionine side chain from the α1 helix and the proton at position 2 of an adenine (Fig. 4H). It was the first time that some dsRBMs were shown recognizing not only the shape of the RNA but also the sequence.[82] More specifically, the structure explains the strong preference for a guanosine moiety 3' to the edited adenosine since dsRBM2 of ADAR2 specifically recognizes the amino group of this base. In interacting with this nucleotide and the one which base-pairs with the editing site, dsRBM2 not only brings the deaminase domain in close proximity to the editing site, but also does not prevent access of the adenosine to the deaminase domain.[82] Finally, this structural study explains how the edited adenine is targeted specifically by ADAR2 among the numerous other adenines located in the stem.

As for ADAR2, sequence specific contacts could also be observed between the $\alpha 1$ helix and $\beta 1$-$\beta 2$ loop of the Aa RNase III dsRBM and the RNA minor grooves.[78,82] However, one difference is that dsRBM of Aa RNase III preferentially recognizes an RNA helix containing a G-X_{10}-G sequence, whereas dsRBM1 and dsRBM2 of ADAR2 bind G-X_9-A and G-X_8-A, respectively. The length and the positioning of the $\alpha 1$ helix relative to the dsRBM fold appear to be the key structural elements that determine the register length of the different dsRBMs.[82] Surprisingly, alignment of several dsRBM sequences reveals a high variability in the length and amino acid sequence composition of the $\alpha 1$ helix and the $\beta 1$-$\beta 2$ loop.[82] In agreement with reports indicating that dsRBMs from different proteins are not functionally interchangeable, it strongly suggests that dsRBMs are likely to have different binding specificities.[87,88]

Conclusion and Perspectives

In this chapter, we have described the current knowledge of how different RBDs interact with RNA at the atomic level and participate in RBP functions. Although still few structures of RBD containing proteins bound to RNA have been determined compared to the vast number of RNA binding proteins, few conclusions or hypotheses can be nevertheless drawn from these structures.

Importantly, the common determinant of the four RBD families described in this chapter is their ability to interact specifically with RNA. Structural biology highly contributed to provide crucial information about this specificity of interaction. For example, it was essential to correctly map binding sites for several splicing factors in vivo (the best examples are Fox-1 and NOVA2), since it revealed that the positioning of these binding sites relative to the splice-sites appears to be a major element controlling the mode of action of these proteins. Although this information is not sufficient to fully characterize this mode of action, it contributes to a better understanding of their functions. It also helped to understand how PTB, U2AF65 and Sex-lethal adapt to the different pyrimidine-tracts found at the 3' splice-site and how Tra2-$\beta 1$ recruits additional splicing factors on the SMN exon7. Finally, solving the structures of RBPs bound to RNA revealed unexpected features like the potential for RNA looping by PTB or MBNL1 suggesting a new function for these proteins in remodelling RNA structure.

However, it is still very hard to predict RBD-RNA interactions due to their versatility of interaction. We showed that the extreme plasticity of RRM for binding RNA can be explained by the use of different combinations of side chain and main chain RNA interactions but also by the capacity for this domain to increase its RNA binding surface outside the canonical β-sheet surface, using an additional β-strand, loops and/or RRM extremities. Zinc fingers use another strategy. Rather than extending their binding surface, they adopt different folds as emphasize in this chapter with the ZRANB2 family. On the contrary KH domains always use the same surface of interaction but still bind different sequences. This variability of RBD-RNA interactions justify the need to determine still more structures of protein-RNA complexes.

Despite progress in the last decade in this growing field, many questions remain to be answered. This ranges from simple questions that could be addressed rapidly by a structural biology approach to more complicated ones that will require multidisciplinary approaches or new methodologies. For example, we still need to address how pseudo-RRMs bind RNA. A more challenging question is how several RBPs assemble or multimerise on RNA? Also, how dynamic are protein-RNA interactions and how posttranslational modifications such as phosphorylation influence this dynamic? Answers to these questions are now needed for a full understanding of posttranscriptional gene regulations.

Acknowledgments

The authors would like to thank the Swiss National Science Foundation (No. 31003AB-133134), the SNF-NCCR Structural Biology and EURASNET for financial support to FHTA and the European Molecular Biology Organization for a postdoctoral fellowship to AC.

References

1. Dreyfuss G, Kim VN, Kataoka N. Messenger-RNA-binding proteins and the messages they carry. Nat Rev Mol Cell Biol 2002; 3:195-205.
2. Lorsch JR. RNA chaperones exist and DEAD box proteins get a life. Cell 2002; 109:797-800.
3. Venter JC, Adams MD, Myers EW et al. The sequence of the human genome. Science 2001; 291:1304-1351.
4. Maris C, Dominguez C, Allain FH. The RNA recognition motif, a plastic RNA-binding platform to regulate posttranscriptional gene expression. FEBS J 2005; 272:2118-2131.
5. Clery A, Blatter M, Allain FH. RNA recognition motifs: boring? Not quite. Curr Opin Struc Biol 2008; 18:290-298.
6. Oubridge C, Ito N, Evans PR et al. Crystal structure at 1.92 A resolution of the RNA-binding domain of the U1A spliceosomal protein complexed with an RNA hairpin. Nature 1994; 372:432-438.
7. Deo RC, Bonanno JB, Sonenberg N et al. Recognition of polyadenylate RNA by the poly(A)-binding protein. Cell 1999; 98:835-845.
8. Handa N, Nureki O, Kurimoto K et al. Structural basis for recognition of the tra mRNA precursor by the Sex-lethal protein. Nature 1999; 398:579-585.
9. Hardin JW, Hu YX, McKay DB. Structure of the RNA binding domain of a DEAD-box helicase bound to its ribosomal RNA target reveals a novel mode of recognition by an RNA recognition motif. J Mol Biol 2010; 402:412-427.
10. Kotik-Kogan O, Valentine ER, Sanfelice D et al. Structural analysis reveals conformational plasticity in the recognition of RNA 3' ends by the human La protein. Structure 2008; 16:852-862.
11. Lunde BM, Horner M, Meinhart A. Structural insights into cis element recognition of nonpolyadenylated RNAs by the Nab3-RRM. Nucleic Acids Research 2011; 39:337-346.
12. Pancevac C, Goldstone DC, Ramos A et al. Structure of the Rna15 RRM-RNA complex reveals the molecular basis of GU specificity in transcriptional 3'-end processing factors. Nucleic Acids Res 2010; 38:3119-3132.
13. Pomeranz Krummel DA, Oubridge C et al. Crystal structure of human spliceosomal U1 snRNP at 5.5 A resolution. Nature 2009; 458:475-480.
14. Price SR, Evans PR, Nagai K. Crystal structure of the spliceosomal U2B"-U2A' protein complex bound to a fragment of U2 small nuclear RNA. Nature 1998; 394:645-650.
15. Sickmier EA, Frato KE, Shen H et al. Structural basis for polypyrimidine tract recognition by the essential pre-mRNA splicing factor U2AF65. Mol Cell 2006; 23:49-59.
16. Teplova M, Song J, Gaw HY et al. Structural insights into RNA recognition by the alternate-splicing regulator CUG-binding protein 1. Structure 2010; 18:1364-1377.
17. Teplova M, Yuan YR, Phan AT et al. Structural basis for recognition and sequestration of UUU(OH) 3' temini of nascent RNA polymerase III transcripts by La, a rheumatic disease autoantigen. Mol Cell 2006; 21:75-85.
18. Wang X, Tanaka Hall TM. Structural basis for recognition of AU-rich element RNA by the HuD protein. Nat Struct Biol 2001; 8:141-145.
19. Yang Q, Coseno M, Gilmartin GM et al. Crystal structure of a human cleavage factor CFI(m)25/CFI(m)68/RNA complex provides an insight into poly(A) site recognition and RNA looping. Structure 2011; 19:368-377.
20. Allain FH, Bouvet P, Dieckmann T et al. Molecular basis of sequence-specific recognition of preribosomal RNA by nucleolin. EMBO J 2000; 19:6870-6881.
21. Allain FH, Howe PW, Neuhaus D et al. Structural basis of the RNA-binding specificity of human U1A protein. EMBO J 1997; 16:5764-5772.
22. Auweter SD, Fasan R, Reymond L et al. Molecular basis of RNA recognition by the human alternative splicing factor Fox-1. EMBO J 2006; 25:163-173.
23. Clery A, Jayne S, Benderska N et al. Molecular basis of purine-rich RNA recognition by the human SR-like protein Tra2-beta1. Nat Struct Mol Biol 2011; 18:443-450.
24. Dominguez C, Fisette JF, Chabot B et al. Structural basis of G-tract recognition and encaging by hnRNP F quasi-RRMs. Nat Struct Mol Biol 2010; 17:853-861.
25. Hargous Y, Hautbergue GM, Tintaru AM et al. Molecular basis of RNA recognition and TAP binding by the SR proteins SRp20 and 9G8. EMBO J 2006; 25:5126-5137.

26. Johansson C, Finger LD, Trantirek L et al. Solution structure of the complex formed by the two N-terminal RNA-binding domains of nucleolin and a pre-rRNA target. J Mol Biol 2004; 337:799-816.

27. Martin-Tumasz S, Reiter NJ, Brow DA et al. Structure and functional implications of a complex containing a segment of U6 RNA bound by a domain of Prp24. RNA 2010; 16:792-804.

28. Oberstrass FC, Auweter SD, Erat M et al. Structure of PTB bound to RNA: specific binding and implications for splicing regulation. Science 2005; 309:2054-2057.

29. Perez-Canadillas JM. Grabbing the message: structural basis of mRNA 3'UTR recognition by Hrp1. EMBO J 2006; 25:3167-3178.

30. Skrisovska L, Bourgeois CF, Stefl R et al. The testis-specific human protein RBMY recognizes RNA through a novel mode of interaction. EMBO Rep 2007; 8:372-379.

31. Tsuda K, Kuwasako K, Takahashi M et al. Structural basis for the sequence-specific RNA-recognition mechanism of human CUG-BP1 RRM3.   Nucleic Acids Res 2009; 37:5151-5166.

32. Tsuda K, Someya T, Kuwasako K et al. Structural basis for the dual RNA-recognition modes of human Tra2-beta RRM. Nucleic Acids Res 2011; 39:1538-1553.

33. Varani L, Gunderson SI, Mattaj IW et al. The NMR structure of the 38 kDa U1A protein—PIE RNA complex reveals the basis of co-operativity in regulation of polyadenylation by human U1A protein. Nat Struct Biol 2000; 7:329-335.

34. Auweter SD, Oberstrass FC, Allain FH. Sequence-specific binding of single-stranded RNA: is there a code for recognition? Nucleic Acids Res 2006; 34:4943-4959.

35. Allain FH, Gubser CC, Howe PW et al. Specificity of ribonucleoprotein interaction determined by RNA folding during complex formulation. Nature 1996; 380:646-650.

36. Jacks A, Babon J, Kelly G et al. Structure of the C-terminal domain of human La protein reveals a novel RNA recognition motif coupled to a helical nuclear retention element. Structure 2003; 11:833-843.

37. Avis JM, Allain FH, Howe PW et al. Solution structure of the N-terminal RNP domain of U1A protein: the role of C-terminal residues in structure stability and RNA binding. J Mol Biol 1996; 257:398-411.

38. Perez Canadillas JM, Varani G. Recognition of GU-rich polyadenylation regulatory elements by human CstF-64 protein. EMBO J 2003; 22:2821-2830.

39. Mazza C, Segref A, Mattaj IW et al. Large-scale induced fit recognition of an m(7)GpppG cap analogue by the human nuclear cap-binding complex. EMBO J 2002; 21:5548-5557.

40. Hofmann Y, Wirth B. hnRNP-G promotes exon 7 inclusion of survival motor neuron (SMN) via direct interaction with Htra2-beta1.   Hum Mol Genet 2002; 11:2037-2049.

41. Young PJ, DiDonato CJ, Hu D et al. SRp30c-dependent stimulation of survival motor neuron (SMN) exon 7 inclusion is facilitated by a direct interaction with hTra2 beta 1.Hum Mol Genet 2002; 11:577-587.

42. Dominguez C, Allain FH. NMR structure of the three quasi RNA recognition motifs (qRRMs) of human hnRNP F and interaction studies with Bcl-x G-tract RNA: a novel mode of RNA recognition. Nucleic Acids Res 2006; 34:3634-3645.

43. Blanchette M, Chabot B. Modulation of exon skipping by high-affinity hnRNP A1-binding sites and by intron elements that repress splice site utilization. EMBO J 1999; 18:1939-1952.

44. Singh R, Valcarcel J, Green MR. Distinct binding specificities and functions of higher eukaryotic polypyrimidine tract-binding proteins. Science 1995; 268:1173-1176.

45. Wolfe SA, Nekludova L, Pabo CO. DNA recognition by Cys2His2 zinc finger proteins. Annu Rev Bioph Biom 2000; 29:183-212.

46. Amarasinghe GK, De Guzman RN, Turner RB et al. NMR structure of the HIV-1 nucleocapsid protein bound to stem-loop SL2 of the psi-RNA packaging signal. Implications for genome recognition. J Mol Biol 2000; 301:491-511.

47. De Guzman RN, Wu ZR, Stalling CC et al. Structure of the HIV-1 nucleocapsid protein bound to the SL3 psi-RNA recognition element. Science 1998; 279:384-388.

48. Dey A, York D, Smalls-Mantey A et al. Composition and sequence-dependent binding of RNA to the nucleocapsid protein of Moloney murine leukemia virus. Biochemistry-US 2005; 44:3735-3744.

49. Zhou J, Bean RL, Vogt VM et al. Solution structure of the Rous sarcoma virus nucleocapsid protein: muPsi RNA packaging signal complex. J Mol Biol 2007; 365:453-467.

50. Hudson BP, Martinez-Yamout MA, Dyson HJ et al. Recognition of the mRNA AU-rich element by the zinc finger domain of TIS11d. Nat Struct Mol Biol 2004; 11:257-264.

51. Lee BM, Xu J, Clarkson BK et al. Induced fit and «lock and key» recognition of 5S RNA by zinc fingers of transcription factor IIIA. J Mol Biol 2006; 357:275-291.

52. Loughlin FE, Mansfield RE, Vaz PM et al. The zinc fingers of the SR-like protein ZRANB2 are single-stranded RNA-binding domains that recognize 5' splice site-like sequences. Proc Nat Acad Sci USA 2009; 106:5581-5586.

53. Lu D, Searles MA, Klug A. Crystal structure of a zinc-finger-RNA complex reveals two modes of molecular recognition. Nature 2003; 426:96-100.

54. Teplova M, Patel DJ. Structural insights into RNA recognition by the alternative-splicing regulator muscleblind-like MBNL1. Nat Struct Mol Biol 2008; 15:1343-1351.
55. Engelke DR, Ng SY, Shastry BS et al. Specific interaction of a purified transcription factor with an internal control region of 5S RNA genes. Cell 1980; 19:717-728.
56. Pelham HR, Brown DD. A specific transcription factor that can bind either the 5S RNA gene or 5S RNA. Proceedings of the National Academy of Sciences of the United States of America 1980; 77:4170-4174.
57. Blackshear PJ. Tristetraprolin and other CCCH tandem zinc-finger proteins in the regulation of mRNA turnover. Biochem Soc T 2002; 30:945-952.
58. Pascual M, Vicente M, Monferrer L et al. The Muscleblind family of proteins: an emerging class of regulators of developmentally programmed alternative splicing. Differentiation 2006; 74:65-80.
59. Plambeck CA, Kwan AH, Adams DJ et al. The structure of the zinc finger domain from human splicing factor ZNF265 fold. J Biol Chem 2003; 278:22805-22811.
60. Nguyen CD, Mansfield RE, Leung W et al. Characterization of a family of RanBP2-type zinc fingers that can recognize single-stranded RNA. J Mol Biol 2011; 407:273-283.
61. Grishin NV. KH domain: one motif, two folds. Nucleic Acids Res 2001; 29:638-643.
62. Valverde R, Edwards L, Regan L. Structure and function of KH domains. FEBS J 2008; 275:2712-2726.
63. Beuth B, Pennell S, Arnvig KB et al. Structure of a Mycobacterium tuberculosis NusA-RNA complex. EMBO J 2005; 24:3576-3587.
64. Braddock DT, Louis JM, Baber JL et al. Structure and dynamics of KH domains from FBP bound to single-stranded DNA. Nature 2002; 415:1051-1056.
65. Du Z, Lee JK, Fenn S et al. X-ray crystallographic and NMR studies of protein-protein and protein-nucleic acid interactions involving the KH domains from human poly(C)-binding protein-2. RNA 2007; 13:1043-1051.
66. Jia MZ, Horita S, Nagata K et al. An archaeal Dim2-like protein, aDim2p, forms a ternary complex with a/eIF2 alpha and the 3' end fragment of 16S rRNA. J Mol Biol 2010; 398:774-785.
67. Lewis HA, Musunuru K, Jensen KB et al. Sequence-specific RNA binding by a Nova KH domain: implications for paraneoplastic disease and the fragile X syndrome. Cell 2000; 100:323-332.
68. Liu Z, Luyten I, Bottomley MJ et al. Structural basis for recognition of the intron branch site RNA by splicing factor 1. Science 2001; 294:1098-1102.
69. Tu C, Zhou X, Tropea JE et al. Structure of ERA in complex with the 3' end of 16S rRNA: implications for ribosome biogenesis. Proc Nat Acad Sci USA 2009; 106:14843-14848.
70. Ule J, Ule A, Spencer J et al. Nova regulates brain-specific splicing to shape the synapse. Nat Genet 2005; 37:844-852.
71. Ule J, Jensen KB, Ruggiu M et al. CLIP identifies Nova-regulated RNA networks in the brain. Science 2003; 302:1212-1215.
72. Ule J, Stefani G, Mele A et al. An RNA map predicting Nova-dependent splicing regulation. Nature 2006; 444:580-586.
73. Berglund JA, Chua K, Abovich N et al. The splicing factor BBP interacts specifically with the pre-mRNA branchpoint sequence UACUAAC. Cell 1997; 89:781-787.
74. Garcia-Mayoral MF, Hollingworth D, Masino L et al. The structure of the C-terminal KH domains of KSRP reveals a noncanonical motif important for mRNA degradation. Structure 2007; 15:485-498.
75. Michelotti GA, Michelotti EF, Pullner A et al. Multiple single-stranded cis elements are associated with activated chromatin of the human c-myc gene in vivo. Mol Cell Biol 1996; 16:2656-2669.
76. Stefl R, Skrisovska L, Allain FH. RNA sequence- and shape-dependent recognition by proteins in the ribonucleoprotein particle. EMBO Rep 2005; 6:33-38.
77. Chang KY, Ramos A. The double-stranded RNA-binding motif, a versatile macromolecular docking platform. FEBS J 2005; 272:2109-2117.
78. Gan J, Tropea JE, Austin BP et al. Structural insight into the mechanism of double-stranded RNA processing by ribonuclease III. Cell 2006; 124:355-366.
79. Leulliot N, Quevillon-Cheruel S, Graille M et al. A new alpha-helical extension promotes RNA binding by the dsRBD of Rnt1p RNAse III. EMBO J 2004; 23:2468-2477.
80. Ramos A, Grunert S, Adams J et al. RNA recognition by a Staufen double-stranded RNA-binding domain. EMBO J 2000; 19:997-1009.
81. Ryter JM, Schultz SC. Molecular basis of double-stranded RNA-protein interactions: structure of a dsRNA-binding domain complexed with dsRNA. EMBO J 1998; 17:7505-7513.
82. Stefl R, Oberstrass FC, Hood JL et al. The solution structure of the ADAR2 dsRBM-RNA complex reveals a sequence-specific readout of the minor groove. Cell 2010; 143:225-237.
83. Yang SW, Chen HY, Yang J et al. Structure of Arabidopsis HYPONASTIC LEAVES1 and its molecular implications for miRNA processing. Structure 2010; 18:594-605.

84. Wu H, Henras A, Chanfreau G et al. Structural basis for recognition of the AGNN tetraloop RNA fold by the double-stranded RNA-binding domain of Rnt1p RNase III. Proc Nat Acad Sci USA 2004; 101:8307-8312.
85. Bass BL. RNA editing by adenosine deaminases that act on RNA. Annu Rev Biochem 2002; 71:817-846.
86. Nishikura K. Editor meets silencer: crosstalk between RNA editing and RNA interference. Nat Rev Mol Cell Biol 2006; 7:919-931.
87. Liu Y, Lei M, Samuel CE. Chimeric double-stranded RNA-specific adenosine deaminase ADAR1 proteins reveal functional selectivity of double-stranded RNA-binding domains from ADAR1 and protein kinase PKR. Proc Nat Acad Sci USA 2000; 97:12541-12546.
88. Parker GS, Maity TS, Bass BL. dsRNA binding properties of RDE-4 and TRBP reflect their distinct roles in RNAi. J Mol Biol 2008; 384:967-979.
89. Ding J, Hayashi MK, Zhang Y et al. Crystal structure of the two-RRM domain of hnRNP A1 (UP1) complexed with single-stranded telomeric DNA. Gene Dev 1999; 13:1102-1115.
90. Koradi R, Billeter M, Wuthrich K. MOLMOL: a program for display and analysis of macromolecular structures. J Mol Graphics 1996; 14:51-55, 29-32.

INDEX

T - #0144 - 111024 - C174 - 229/152/8 - PB - 9780367445911 - Gloss Lamination